Computers and Mathematics

The Use of Computers in Undergraduate Instruction

A project of the
Committee on Computers in Mathematics Education
of The Mathematical Association of America

Editors

David A. Smith, Duke University
Gerald J. Porter, University of Pennsylvania
L. Carl Leinbach, Gettysburg College
Ronald H. Wenger, University of Delaware

Acknowledgement

The Mathematical Association of America and its Committee on Computers in Mathematics Education express their appreciation for the support of the IBM Corporation for this project. The generosity of IBM has covered the added cost of phototypesetting over laser–quality printing and the cost of providing copies *gratis* to mathematics departments in North America.

We also express our thanks to TCI Software Research, Inc., whose version 2.2 of their T^3 technical word processor and whose labor have captured and set down this text. For details see the colophon at the end of this book.

The Committee also expresses its thanks to Duke University for the secretarial and other support that allowed David A. Smith to orchestrate most of the production of these notes.

Library of Congress Catalog
Card Number 87–063425
ISBN 0–88385–059–1

Table of Contents

INTRODUCTION

A Message from CCIME

The MAA Committee on Computers in Mathematics Education (CCIME) is presently composed of the editors of this volume and Chris W. Avery (DeAnza College), Sheldon P. Gordon (Suffolk County Community College), Eugene A. Herman (Grinnell College), Zaven A. Karian (Denison University), Howard L. Penn (U. S. Naval Academy), Alan H. Schoenfeld (University of California, Berkeley), and J. Arthur Seebach (St. Olaf College). Our mailing list also includes some two dozen "friends" of the Committee, many of whom participate more or less regularly in its meetings and activities. We meet at every national meeting of MAA and otherwise conduct business by mail, phone, or electronic mail. Anyone interested in the work of the Committee is welcome to attend meetings, to volunteer assistance, and/or to be on the mailing list; just send name, address, phone number, and electronic address (if any) to the chairman.

Over the past three years the most visible activity of CCIME has been the sponsorship of panel discussions, contributed paper sessions, and an hour speaker (only one, so far) at summer and winter national meetings. This volume is in part an outgrowth of a desire to expand our audience to include those not able to attend national meetings and those who have missed our "live" presentations.

In particular, several of the papers presented here are based directly on or are updates of CCIME panel discussions. To round out the coverage of topics we thought important to include, we invited other knowledgeable individuals to write articles specifically for this volume. Not all of our invitees were able to contribute, but we think our coverage is representative of present practice in the use of computers in undergraduate mathematics instruction.

We never intended a thorough survey of all computer use in mathematics, nor did we attempt to identify every individual working at the cutting edge. What we offer is a range of experiences, ideas, predictions, useful information, and pointers to other sources of information that we think most college–level mathematics instructors will find helpful.

Jerry Porter's Preface establishes the historical context for the volume and makes some predictions of future developments. Alan Schoenfeld's lead article provides a philosophical basis and gives specific examples of some of the more exciting developments already available or soon to be. The "general" section of the volume also includes Steve Cunningham's practical advice on how to assess the value of software for a particular educational application.

The remainder of the volume is in some measure "course specific." Carl Leinbach surveys past and present uses of computers in calculus. The following papers by David Smith and John Hosack lay out opposite ends of a technological spectrum of possibilities for using a computer as a tool in undergraduate courses; neither is limited to a single course, but both draw on calculus for examples. Martha Siegel describes some possibilities for computer use in the discrete math courses being offered at many schools prior to, in parallel with, or in place of calculus. Jim Sandefur and Andrew Vogt describe an alternate course that would hardly be possible at the freshman level without computers: finite difference calculus, with an emphasis on discrete dynamical systems.

Gene Herman's linear algebra panel discussion, which includes contributions by Howard Anton, Alan Tucker, and Garry Helzer, is presented almost as it happened at the Laramie meeting in 1985. It is supplemented by an article by Morris Orzech on the same subject. Each of the five authors has his own approach to computer use in linear algebra, and yet there is an intriguing common message in all five presentations.

The coverage of differential equations is based on a panel discussion presented this year in San Antonio. It includes a selection of organizer Howard Penn's reviews of current software and an expanded version of Tony Danby's presentation on the use of applications in the "standard" ODE course.

The papers by Florence Gordon on statistics and by Laurie Snell and John Finn on probability survey a wide range of ideas, with small but non–empty intersection, for computer use in these courses.

The next two papers have a common thread in the use of Logo (also touched on by several other authors, but still relatively rare at the college level), yet they describe very different courses. Charles Jones has developed a "math for poets" liberal arts course whose central theme is problem solving. James King makes a compelling argument for a renaissance of geometry in the curriculum by showing how relatively simple explorations with Logo can lead to discovery of deep mathematical concepts.

Chris Nevison's paper on abstract algebra and number theory surveys developments in those areas and updates a panel discussion he organized for the Anaheim meeting in 1985.

The last two papers (surely not "least" by several different measures) deal with courses considered "remedial" or "transitional" or "developmental" in the college curriculum. Gloria Gilmer and Shelly Gordon report on an informal survey of individuals at a wide range of institutions concerning the ways in which they use computers in such courses and the effectiveness of doing so. Ron Wenger describes in some detail the multifaceted program at the University of Delaware and offers some insights on the future prospects for such programs.

One theme that runs like a thread through all the papers in the volume is the use of the computer as a *tool* rather than as an object of study in its own right. All of our authors focus clearly on teaching and learning *mathematics* to the near exclusion of matters we might classify as computer science or computer engineering. We don't have to know how to design or build a car in order to find one useful, and the business world has shown us that the same is true of computers.

The work of many individuals is represented here. The project was started under the leadership of the first chairman of CCIME, Jerry Porter. The contributions of authors and editors are explicitly acknowledged as such, and several of the papers contain acknowledgements to other individuals. We are especially indebted to **TCI Software Research, Inc.**, developers and publishers of the **T^3** technical word processing system. As a demonstration of the capabilities of their new version 2.2 of **T^3**, they have graciously offered to print the master copy of this volume at no cost to MAA, which has enabled us to keep the cost of the book low and still have a uniform and attractive format. We also thank the secretarial staff of the Duke University Department of Mathematics for retyping a number of the papers in **T^3**. Some of the contributions were submitted on disks or via Bitnet from a variety of other word processors, and all of those were imported directly into **T^3** for final editing.

David A. Smith
Chairman, CCIME
August, 1987

PREFACE

The Use of Computers in Mathematics Instruction Past History — Future Prospects

Gerald J. Porter

The digital computer was created to facilitate the solution of problems that required computations which were, for all practical purposes, impossible using previous technologies. Subsequent developments of mainframe computing have been similarly motivated either by the need to solve increasingly complex scientific problems or by the need to manipulate huge quantities of information. Digital computers first made their appearance on college campuses in the late 1950s. During the next decade nearly every college and university acquired access to computing resources primarily for administrative and scientific/engineering type applications.

Although computing equipment was not acquired for instructional purposes, the researchers who were using computers quickly realized that the computer could be used to illustrate and motivate concepts which were difficult to explain otherwise. Virtually all computing was done in batch mode during these years and programming was done using a scientific programing language such as FORTRAN. Typical instructional use followed this pattern:

a) a student would write a program;

b) the student would journey to a computer center where he or she would punch the program onto specially formatted "computer cards";

c) the cards would be submitted to a computer center dispatcher, who would enter the job into a job queue;

d) some time thereafter, dependent upon the computer job load, the computer would attempt to execute the program submitted;

e) under ideal circumstances, the program would run and produce the desired output, which would be gathered by the dispatcher and placed in an output bin for the student;

f) more likely, the student had mispunched a card or violated the stringent syntactical rules of FORTRAN, and the computer simply identified one or more syntax errors, in which case the student returned to step b) above;

g) even when there were no syntax errors, there still might be programming errors, i.e. the program ran but it did not do what it was intended to do, in which case the student returned to step a) above.

Under the best conditions, the time from submission to return of the job would be 45 minutes. If, perchance, the assignment was submitted at the time that the school's payroll needed to be processed, the turn–around time was greatly increased.

This computer environment required that an account for each student be set up at the computer center, that the student learn a programming language, and that a maximum of one assignment per week be given. Despite these obstacles, computing was introduced into mathematics courses, primarily into calculus and differential equation courses. Typical uses included numerical calculation of limits, derivatives, and integrals. It is fair to describe this period by saying that a numerical analysis component was introduced into calculus. This mode of instruction is described more fully in Carl Leinbach's article in this volume. Most of the assignments from that period can easily be done today on a hand held programmable calculator smaller in size than the deck of computer cards required then for a single assignment.

The unforgiving nature of FORTRAN syntax and the complexity of its input–output statements required that a more "user friendly" language be created. There was agreement that an hour of class time spent on FORMAT statements could be spent better on other topics. As a result, alternative languages such as BASIC and WATFOR were created. These languages, still used in the batch mode, enabled students to complete their assignments more quickly, since fewer runs were required. In addition, they shortened the initial period required to learn programming.

The next major development in computing which affected instructional use was interactive computing. Under this mode of operation, a user could connect with a mainframe from a remote location, enter a program, and execute it. The computer operating system was set up so that the computer would run many jobs "simultaneously." In addition, stored programs could be directly accessed from the remote terminal.

New applications of the computer to instruction became possible. Perhaps the major new development was computer aided instruction (CAI). In its simplest form CAI consisted of drill sessions in which a student would be presented with multiple choice questions. The basic idea was: 1) student learning would be reinforced by correct answers, and 2) students would be able to identify those areas in which their knowledge was inadequate. In addition, instructors would be able, by keeping track of student records, to determine topics in which further training was required. Such applications were used primarily in precalculus courses. For a further discussion of this mode of instruction, see the article by Gloria Gilmer and Sheldon Gordon in this volume.

More generally, interactive computing allowed instructors to prepare instructional modules which students could modify and run without having to write programs. For example, a program could be made available to students to illustrate the convergence of rectangular approximations to the definite integral. The student would enter the function, the limits, and the step size. By repeatedly decreasing the size of the subdivision, the student could see that rectangular approximations to the definite integral appear to converge numerically.

The output from most of the programs at this time was numerical or perhaps consisted of "printer graphics," i.e., symbols printed in the standard line and character matrix of the line printer. Some instructors, however, began to use graphics terminals (primarily utilizing storage tube technology) to illustrate such concepts as the convergence of Taylor polynomials to a given function. The use of graphics was severely limited by the technology (individual points or lines could not be erased) and the cost of graphics terminals. In addition, these devices were usually connected to remote mainframes via telephone lines at 300 baud. Consequently, pictures were drawn very slowly (line speed) and in bursts (time sharing).

With the introduction of the microcomputer in the late 1970s, graphics became widely available. Graphical presentations in color with a reasonable resolution were now available on standalone microcomputers costing a few thousand dollars. Limitations of baud rates and storage tubes became a thing of the past. Distributed data and program storage became possible through the use relatively inexpensive floppy disks. Perhaps more importantly, the microcomputer changed the locus of control from the campus computer center running large mainframes in a multiuser environment to a departmental computer laboratory controlled by the instructional staff and available without the need to secure budgets and create student accounts. Many of the previous applications, including CAI and graphics, were now transported to micros, which became the computers of choice for mathematics instruction.

With the introduction of the IBM personal computer and the Apple Macintosh, bit mapped screens became available for sophisticated graphics use, the amount of random access memory increased from the first generation's 48K to 512K or more, and the processors ran more quickly. As a result programs of increasing sophistication could now be run on a micro. Examples of such programs include PC–MATLAB, a library of mathematical programs, and muMath, a symbolic algebra program. (See John Hosack's article in this volume for additional discussion of computer algebra systems.) Not only did microcomputers become more sophisticated, but their prices declined to the point where some schools could require that students purchase their own microcomputers, and textbooks are now sold with disks of computer programs. Again, the locus of control changed, this time from the instructor to the learner.

The declining price of computers has made it possible for most high schools to have computers available for their students. The College Board's advanced placement examination in computer science has provided another incentive for secondary schools to teach computer skills. The number of college freshman taking Introductory Computer Science has dropped sharply over the past few years due to the fact that many students acquire these skills in high school.

The development I have sketched above is one of an ever changing technology driven primarily by non–instructional needs. Nevertheless, as technology changes the instructional applications of technology also change. As we look to the future, the one thing we can be sure of is that technology will continue to advance and, on the whole, will continue to decline in price.

Thus we find ourselves in 1987 in an environment in which our students arrive at college knowing how to use a computer and able to purchase a microcomputer for approximately a thousand dollars with a capability greater than many of the mainframes of 1967. For less than a hundred dollars hand held calculators are available with programming and graphics capability equivalent to microcomputers of 1978. There can be little doubt that this technology will affect both the way we teach undergraduate mathematics and the content of our courses. As we look to the future, we see an ever changing technology which will continue to provide opportunities for instructional innovation. The advanced workstations of today will become the commonplace equipment of tomorrow. Projects such as Andrew at CMU, Athena at MIT, and Iris at Brown are aimed at an environment in which a high performance workstation with high resolution graphics, expanded memory, and fast processors is available to each student. These machines will be linked in networks which will access file servers, high quality printers, and gateways to external resources such as remote data bases and libraries.

Instead of focusing on the use of computers to improve the teaching of current mathematical topics, we can begin to explore simulated learning environments which will enable us to teach materials previously thought to be too abstract or complex for undergraduates. Thomas Banchoff's work at Brown University in teaching differential geometry is an example of work already begun in this direction.

Storage devices such as CDROM's will make possible the distribution of machine readable texts much larger than today's textbooks, which need not be accessed in a linear manner. Continual student testing, done interactively, can determine the level of presentation and the necessity for review. Homework assignments can easily be generated which will include review questions, as needed, as well as exercises on the current material. For those students who learn better from oral presentations than from texts, intelligent videodisks can be utilized to provide lectures and "problem sessions" at a level appropriate for the individual learner.

The continued transfer of today's microcomputer technology to tomorrow's hand held calculators will lead to hand held devices for computer algebra systems and matrix calculations. The widespread availability of such devices can have a profound effect in transferring the focus of undergraduate instruction away from training students to perform relatively routine algorithms and toward the theory behind these algorithms and more sophisticated applications. Current work in artificial intelligence and expert systems may provide intelligent tutors capable of diagnosing and helping to correct learning problems in mathematics. A fundamental prerequisite for such systems is an improved knowledge of the ways in which individuals learn mathematics. Research on this topic is a focus of activity in cognitive science.

The National Science Foundation's supercomputing initiative has enabled researchers to utilize vector and parallel processing. These activities will have an effect both on research in numerical methods and on the content of undergraduate instruction on these topics. It is difficult at this time to predict the effect of vector processing on the way we teach undergraduate mathematics; however, if past behavior is a clue to the future, there can be little doubt that researchers will apply their research tools in their instructional activities.

Twenty years ago an NSF Conference on Computers in Undergraduate Education concluded[1]:

> "The availability of computers has created new challenges for curriculum development in mathematics. Specifically, the undergraduate mathematics curriculum must be changed to take into account new ways of solving old problems and the host of new problems which have arisen. Students will be enormously stimulated by the introduction of computing into the curriculum; for it will turn the all–too–often totally passive college experience into one of active participation.
>
> "If the basic undergraduate mathematics courses are not appropriately modified to reflect the points of view which are associated with computer applications in mathematics, these course will lose much of their relevance for the coming generation of college and university students."

Since that time, significant advances have been made in computer technology, yet computers are still not widely used in undergraduate mathematics instruction. There are many reasons why this is the case. Initially, computers were "unfriendly" objects and a chore to use; this is no longer true. Initially, our colleagues were not, by and large, computer literate; this too has changed. Initially, few sharable materials were available; now there are large data bases of software for use in mathematics instruction. What has not changed, however, is the undergraduate mathematics curriculum. Undergraduate mathematics is taught today more or less the same way it was taught twenty years ago. Attempts to superimpose computer assisted exercises on a curriculum which evolved in an environment without significant computational resources have proved to be futile. Our colleagues twenty years ago identified the work to be done with great precision: **The undergraduate mathematics curriculum must be changed to take into account the availability of significant computational facilities.** That task remains to be completed. Without fundamental changes in the mathematics curriculum, computing will not be utilized to the extent that many of us feel is appropriate and desirable.

This volume is intended to document the progress we have made in using computers to improve undergraduate mathematics instruction. It is the editors' hope that the work chronicled here will serve as a foundation for continued activity, which will bring us closer to the realization of the opportunities technology offers for the improvement of undergraduate mathematics instruction.

---

[1]**Proceedings of a Conference on Computers in Undergraduate Education: Mathematics, Physics, Statistics and Chemistry**, sponsored by the National Science Foundation, conducted at The Science Teaching Center of the University of Maryland, December 8–9, 1967.

Contributors' Addresses

Howard Anton
304 Fries Lane
Cherry Hill, NJ 08003

R. Stephen Cunningham
Department of Computer Science
California State University,
Stanislaus
Turlock, CA 95380

J. M. A. Danby
Department of Mathematics
North Carolina State University
Raleigh, NC 27695

John Finn
Department of Mathematics
and Computer Science
Dartmouth College
Hanover, NH 03755

Gloria Gilmer
The Math–Tech Connexion, Inc.
2014 W. McKinley Ave.
Milwaukee, WI 53205

Florence S. Gordon
Department of Mathematics
New York Institute of Technology
Old Westbury, NY 11568

Sheldon P. Gordon
Department of Mathematics
Suffolk County Community College
533 College Road
Selden, NY 11784

Garry A. Helzer
Department of Mathematics
University of Maryland
College Park, MD 20742

Eugene A. Herman
Department of Mathematics
Grinnell College
Grinnell, IA 50112

John Hosack
Department of Mathematics
Colby College
Waterville, ME 04901

Charles A. Jones
Department of Mathematics
Grinnell College
Grinnell, IA 50112

James R. King
Department of Mathematics
University of Washington
Seattle, WA 98195

L. Carl Leinbach
Department of Computer Studies
Gettysburg College
Gettysburg, PA 17325

Christopher H. Nevison
Department of Computer Science
Colgate University
Hamilton, NY 13346

Morris Orzech
Department of Mathematics
and Statistics
Queen's University
Kingston, Ontario
Canada K7L 3N6

Howard Lewis Penn
Department of Mathematics
U. S. Naval Academy
Annapolis, MD 21402

Gerald J. Porter
Department of Mathematics
University of Pennsylvania
Philadelphia, PA 19104

James T. Sandefur
Department of Mathematics
Georgetown University
Washington, DC 20057

Alan H. Schoenfeld
School of Education
University of California
Berkeley, CA 94720

Martha J. Siegel
Department of Mathematics
Towson State University
Towson, MD 21204

David A. Smith
Department of Mathematics
Duke University
Durham, NC 27706

J. Laurie Snell
Department of Mathematics
and Computer Science
Dartmouth College
Hanover, NH 03755

Alan C. Tucker
Department of Applied Mathematics
and Statistics
State University of New York
at Stony Brook
Stony Brook, NY 11794

Andrew Vogt
Department of Mathematics
Georgetown University
Washington, DC 20057

Ronald H. Wenger
Mathematical Sciences Teaching
and Learning Center
032 Purnell Hall
University of Delaware
Newark, DE 19711

Uses of Computers In Mathematics Instruction

Alan H. Schoenfeld

People have noted, with irony, that we mathematicians are the most computer–phobic of scientists; most of us stick to pencil and paper, while our colleagues in various scientific disciplines have come to rely on high–powered computational tools and techniques. (How many mathematics faculty, faced with a tough differential equation or a complex system of linear equations, head for their terminals to access solutions on muMath or MACSYMA, rather than crank them out longhand? The solution via technology is both faster and more likely to be correct.) If you expand the notion of computer use to include tools that make life easier, such as text processors, mathematics faculty may be among the most computer–phobic of college and university faculty. Exponentially increasing numbers of professors of English and Linguistics are producing and revising their papers on line (either on mainframes or on micros), but one still has the image of the archetypal mathematician scrawling a handwritten manuscript and giving it to an overworked secretary who turns it around in a couple of weeks, only to have it corrected, revised, and re–typed (with new typos) in an iterative process that eventually converges on a final version of a paper. To put things simply, the vast majority of us haven't yet become comfortable with the idea of computer as tool.

As a community, we are even less familiar with computers as tools for teaching than we are with computers as tools in our professional lives. We have only begun to explore the potential uses of computer–based technologies for educational purposes. Indeed, this problem is far more severe at the college level than it is at the elementary and secondary levels. Since the subject matter we teach is substantially more difficult, much more work and sophistication has to go into building systems that will be of use to our students. (It's one thing to build a drill–and–practice program or a computer–based tutor for work on the quadratic equation. It's quite another to do the same for max–min problems!)

The entire educational community — all levels, all subjects — is just beginning to make use of computers in instruction. As Seymour Papert notes in *Mindstorms* [12], the first uses of new tools are usually rather constrained; people use the tools to do things they did before, maybe a bit better or faster. Papert argues that we have barely begun to imagine the ways that computer–based technologies might be harnessed for education, in or out of the classroom. He and others argue that computers are one of a rare breed, a transcendent technology — a technology that has the potential to cause significant change. Of course, we must be wary of such claims; recall the claims two decades ago that educational television would revolutionize schooling in America. But we should be equally wary of saying that "computers have had little significant impact in the schools, so they are likely to play the same minor role as educational television." As Andy diSessa notes, in medieval times you'd have been likely to claim that this newfangled thing called "literacy" wasn't going to have a significant impact — if your extrapolations were based on observations of people who had access to books once a week, for fifteen minutes at a crack.

In this essay I would like to give a sense of what is possible in the near–term future, rather than a discussion of what is known to work right now. My comments will be based upon, but not limited to, current practice; I shall rely on the other essays in this volume to give a good sense of the state of the art. My intention is to outline different "modes of use" for computers in education, illustrating how technology might be harnessed in various ways for educational purposes. Since few mathematical exemplars exist — especially at the college level — I'll use examples from a wide range of disciplines. Those examples serve as categorical indicators of things we might hope to develop for ourselves. (For more extended discussions of this and related issues in mathematics instruction, see [13] and [14]. For a discussion of technology in the future of education, see [11].)

We begin with a discussion of some of the more familiar and straightforward uses of computers for educational purposes. We then proceed to some of the less known and more visionary perspectives on the topic. The following types of computer–based environments and modes of use will be discussed, in this order: drill–and–practice; tools that do the drudgework, so you can use your brains; gaming environments and multiple representations; simulations; dynamic representations; non–standard consequences of learning to program;

intelligent tutoring systems; computer–based microworlds; and transcendent technologies and general all–purpose mathematical tools.

Computer–based drill–and–practice.

I must confess that this mode of use isn't terribly exciting (and it won't get much space here), but it's useful in some contexts and should be mentioned. Some skills have to be mastered, and mastering them is a matter of practice and feedback. Even the most rudimentary computer–based systems can offer significant advantages over paper–and–pencil drill and practice:

Immediate feedback. When working problems using pencil and paper, a student with a misconception can get lots of problems wrong before discovering that he or she has a problem. When using an on–line system, the student can get immediate feedback. When the student makes a mistake, it's brought to his or her attention. This can nip a problem in the bud, keep the student from wasting lots of time and energy, and prevent bad habits from becoming engrained.

Problem sequencing. With written problem sheets, the sequence of problems a student works is determined in advance. In some cases, the student might profit from working more easy problems before going on; in others, the student will be bored stiff by repetitive exercises testing things that have already been mastered. This kind of problem is easily avoided with computer–based record–keeping and problem generating systems. Problems can be characterized according to "level of difficulty." Students can begin work at level 1, and then be allowed to move up to level $n+1$ when they've demonstrated mastery of level n — perhaps by getting five problems right in a row or meeting some other criterion of success. The computer can also keep track of individual students' histories so that each student can re–enter the system where he or she left off.

Motivation and other unexpected benefits. One example of an utterly boring drill–and–practice task, at the upper elementary school level, is solving ratio and proportion problems such as

$$\frac{3}{7} = \frac{X}{21}.$$

All such problems are solved by a basic two–step algorithm: cross–multiply, then divide both sides of the resulting equation by the number that's multiplied by X. It's hardly interesting, either in terms of the task or in terms of the thinking involved. Yet, according to Hugh Burkhardt, a trivial practice program based on this task produced some interesting side effects.

The program was set to generate one new ratio problem every few seconds, with the problem display filling up most of the screen so that a whole class could see it without difficulty. With the set turned on, the students saw

$$\frac{4}{?} = \frac{6}{9}$$

for a brief while. The answer flashed on the screen, and then a new problem, say

$$\frac{1}{?} = \frac{5}{10},$$

appeared. Two very interesting things happened when this game was used in the schools. First, kids took the questions as a challenge and made the display part of a contest: Who could answer fastest? Burkhardt reports that when the students went into the schoolyard for recess and noted that the monitor, which had been left on, was visible from outside, a number of them bunched around the classroom window and continued to play the game. So, via technology, something dull and boring became something interesting. Second, playing the game pointed out something for teachers to think about. One tends to think of the solution to ratio problems in terms of the two–step algorithm mentioned above: cross–multiply, then divide. But when you play the game you don't do it that way at all. You look for relationships among the given integers — sometimes it's across, sometimes it's down, and sometimes it's not immediately apparent. In these different circumstances you do different things. What that means from the teacher's point of view is that you should teach these different procedures. The formal procedure may be all that's necessary in theory, but developing an understanding of the ratio means more than that. In sum, this trivial game served as a stimulus for thinking about how you **do** mathematics, as opposed to how you characterize it.

Tools that do the drudgework so that you can use your brains.

There are at least two different kinds of tools in this category: (a) tools that do computations for students, in effect freeing them to focus on conceptual ideas, and (b) tools you can "program" as aids to making conjectures. (The boundaries between the two are not firm.)

There is nothing special about the tools in category (a), in that virtually any computer–based number– or symbol–cruncher useful to mathematicians can also be exploited to help students. What matters here is the way that the tool is exploited. Consider, for example, symbolic manipulation packages such as muMath or MACSYMA. Once you have gotten past the "overhead" in learning to use the systems, you can use them to do hackwork computations, e.g., differentiating and integrating, solving equations of the form $f(x) = 0$, solving systems of equations, doing matrix manipulations, etc. With such tools available, classroom examples need not be limited to easily solved equations with integer solutions. Since real–world problems rarely come in that form, using the computational tools enables us to tackle much more realistic problems in our "applied" classes. But it does more. By doing the hackwork for students, it lets them focus on which questions they should be asking.

For example, one colleague tells the story of what he did in a calculus class where the students were learning to graph rational functions. Typically, the hardest functions you'll deal with in such classes are of the form $f(x) = P(x)/Q(x)$, where P and Q are cubics. The class warmed up on these, discovering some basic properties of such functions. Once they had the basics down, they moved to a classroom with a terminal. The teacher wrote a ridiculously complicated rational function on the board — P and Q of degree sixteen and seventeen respectively, and not neatly factorable — and posed the task for the students: Figure out what the graph of the function looked like. They could ask him any computational question they liked [e.g., What is $f(2)$ or $f(2000)$? Solve $P(x) = 0$. Compute $f'(x)$], and he'd give the answer. But they had to figure out which questions to ask. The students' initial attempts were clumsy. There was some fumbling at the beginning, as the students identified some major features of the graph, such as zeros, and vertical asymptotes. Then, slowly, the students made sense of the behavior of the function between its first two vertical asymptotes — mostly by picking evenly spaced values and seeing how the function behaved. They were a bit faster between the second and third asymptotes, having noticed that focusing on the function's behavior near the asymptotes was more informative than using even spacings. And as the students made their way across the board, they got better and better at asking the "right" questions. That is, they learned to focus on what really makes the function tick. They did so because they could focus on the function's behavior; they weren't all tied up doing the calculations. One can easily imagine using computer–based tools similarly in a wide range of mathematics courses, pure and applied. (For example, there are the beginnings of a revolution in the secondary school statistics curriculum. With software packages to do number crunching, students can do honest–to–goodness exploratory data analysis. See, for example, [16].)

My second example is a tool specifically designed to engage students in the give–and–take of mathematical invention and conjecture. In most high school geometry classes, students memorize the proofs of required theorems and some standard constructions. Occasionally they get involved in a particular problem, e.g., trying to trisect an angle with straightedge and compass. (I spent quite a bit of time on that one when my teacher said that I wouldn't be able to do it. She was right.) But extended investigations aren't too likely to take place, partly because the cost of testing a conjecture — in time and energy — is so large. If you had a construction that was 43 steps long, would you want to try it out on a dozen different figures?

The "Geometric Supposers," developed at the Harvard Educational Technology Center, are computer–based slaves that will do your constructions for you [17]. You specify the procedure the first time, e.g.: "Give me a triangle ABC. Bisect AB, and label the midpoint of AB as D. Bisect BC, and label the midpoint of AB as E ... ." As you type in the instructions, the constructions you have defined appear on the screen. But now that you've defined you construction, you have the whole construction accessible as a procedure. You can ask for a new triangle ABC, and the Supposer will repeat your construction. This permits you to test conjectures quickly on new figures. The "overhead" of performing excruciatingly detailed constructions with straightedge and compass is no longer an obstacle to trying out ideas.

Of course, these explorations have to take place in an appropriately designed context. In the wrong context, one can imagine students becoming rampant empiricists: "We don't need proof at all, because we can test any idea on hundreds of examples. If it works, it must be right." But in the right context, things can go the other way. Judah Schwartz tells me that students come to disbelieve empirical proof: "So what if you've gotten it to work on ten examples. We've seen constructions that worked on thirty and turned out to be wrong." But perhaps more important, the presence of the Supposers tends to change the nature of classroom inquiry. With these powerful construction tools, students come to see patterns and make conjectures that go far beyond standard course content. (Suppose you go 1/3 of the way down AB to point D and draw CD, then go 1/3 of the way down CB to point E and draw AE, then label the point of intersection of CD and AE as the point F. It sure looks like the extension of BF passes through the midpoint of AC. Can that possibly be right?) In fact, the conjectures may go beyond what the teacher knows is true — and when that happens, teacher and student can do mathematics together. That's quite a bit different from standard practice, where the teacher passes on an established body of knowledge for the students to "master."

Gaming environments and multiple representations.

Graphs are one representation of (some) functions, algebraic formulas another. Thanks to Descartes and those who elaborated on the Cartesian theme, we see the linkages between them. Much of the high school curriculum is devoted to making this connection, often through laborious plotting of graphs on a point–by–point basis. Because so much labor goes into doing the graphing, it takes a long time for students to "rise above" the graphs to see their properties, to recognize that the algebraic form of an equation often tells you about the geometric form of the graph, and to see what the connections between the algebraic and geometric forms are.

A piece of software called "Green Globs" ([8], [9]) helps students see those connections. One of the games in Green Globs (there are four) works as follows. The game board is a subset of the coordinate plane ($-10 \leq x \leq 10$, $-8 \leq y \leq 8$, with the x– and y– axes clearly marked). When the game starts, thirteen randomly generated "globs" (circles of radius approximately 1/2) appear on the screen. The globs are targets. Students shoot at the globs by typing the equation of a curve. The curve is drawn, and if the curve passes through any of the globs on the screen, the globs explode and the student scores some points. The number of points scored grows exponentially with the number of points hit, so there is a clear incentive for hitting as many globs as possible with one shot.

In this game, the task is not to plot $y = f(x)$ or $g(x,y) = 0$ — the standard academic task. Rather, it's to say "I want to find a function $y = f(x)$, or $g(x,y) = 0$, that is shaped a certain way." Suppose, for example, a student is trying to hit three globs with a parabola, $y = ax^2 + bx + c$. The first shot quite often doesn't work — say, the parabola is too wide, or it goes below the points the student is shooting for. The student's task is then to modify his shot: "How can I change the equation of this parabola to make it narrower (in the first case) or move it upward (in the second)?" By making the new shots and seeing the results, students learn the effects of varying the parameters a, b, and c on the graph of the parabola.

In software under development at Harvard's Educational Technology Center, there is a "multiple representations world" that shows three items simultaneously. First, it shows a set of icons representing a relationship. For example, for the statement "there are three boys for every four girls," part of the screen shows a number of boxes, each of which contains three boys and four girls. Students can add to the collection of boxes one at a time, collecting first three boys and four girls, then six boys and eight girls, then nine boys and twelve girls, and so on. As they do, entries are added to a (boys vs. girls) table, and points are plotted on a (boys,girls) set of axes. The result — the developers hope — is that students will come to see that all of these representations are characterizations of the same thing.

These two examples are rather elementary, and it takes a leap of imagination to generate something similar for collegiate mathematics. To my knowledge nothing comparable exists — but one can imagine an "algebra world," where two isomorphic groups are manipulated in such ways that it is apparent that the group operations are the same.

Simulations.

Before they are allowed to pilot certain planes for real, some jet pilots are required to practice in a flight simulator — a computer–driven environment in which the pilot sits in a realistic cockpit and is put through training exercises. The pilot manipulates the cockpit controls, and the computer produces the reactions (the panel displays and the motion of the cockpit) simulating those that the pilot would experience if he were actually in flight. By virtue of this simulated experience, student pilots train in safe circumstances for experiences that would be quite hazardous if they did them in real circumstances.

Some computer–based simulations of various physical systems are a little more down–to–earth, for example, those of steam generation plants [10] and of complex electronic circuits [5]. In the former case there are parallels to the flight simulator: Individuals make decisions (to close a valve or a steam chamber when the reading on a gauge gets too high, for example) and see the results (a return to safe conditions or perhaps an unexpected side effect). But there are also some differences, because in some significant ways the simulation goes beyond the real world experience. In the plant control room, you see only gauges, valves, and pipes; you make a decision, and then get feedback from the gauges as to whether things are going the way you hope they will. Over long experience with the plant, you build a "mental model" of what happens in various parts of the plant when you manipulate the controls. But in the simulation, all of those connections are made explicit, and you get to see the results of your actions: If a decision to close one valve would result in a particular chamber overheating, you see the chain of connections explicitly on the screen. The idea is that such explicit linkages help you build your mental model of the plant before you take control.

The electronic circuit simulations work in similar ways. In some systems, simulated meters appear at all the components of a circuit, giving you a full set of readings. When you change something in the circuit — say, vary a rheostat, replace one resistor with another, or insert a new component into the circuit — the meter readings change automatically. As a result, you can trace the consequences of your actions much more easily than you can while working on real circuits. In some more sophisticated systems (see the discussion of intelligent tutoring systems below), students are presented with schematic diagrams and simulations of faulty circuits. They can request readings, and their "debugging strategy" can be compared with an expert's for purposes of tutoring.

In sum, simulations allow people to see (representations of) entities that may be invisible, or difficult to see, when working on "real world" problems. Clearly such simulations are useful in applied mathematics. It remains to be seen whether we can create analogous simulations for use in the undergraduate curriculum.

Dynamic representations.

Like simulations, dynamic representations make it easier to understand complex phenomena by making representations of them tangible and visible, often in "real time." I give two examples here.

Students have significant difficulty understanding the relationships among distance, velocity, and acceleration. For example, if we give calculus students a velocity–*vs.*–time graph and ask them to produce a distance–*vs.*–time graph from it (or *vice versa*), the odds are high that absolute chaos will result. Some software developed by Bob Tinker deals with this problem head–on. Equipment attached to a microcomputer measures a person's distance from a sensor as a function of time. This allows a student to generate a distance–*vs.*–time graph by walking back and forth from the sensor. In fact, the program will graph distance, velocity, or acceleration *versus* time (one by one or simultaneously, in real–time or in playback). In a typical use of the program, one student generates a graph, say, an acceleration–*vs.*–time graph. Then another student tries to replicate the graph. Moving back and forth from the sensor, the student tries (in real–time) to match acceleration — learning in the process that, for example, you can still be moving forward when your acceleration is negative.

A second example comes from my current work. As part of an intelligent tutoring system (see below) for functions and graphing, my research team at Berkeley has built a "graphing tool kit" with which students can investigate the relationship between the equations and graphs of algebraic functions. Motivated in part by the intuitions behind "Green Globs," we wanted students to be able to see how varying the parameters in algebraic expressions — the boldface letters in symbolic forms such as $y = \mathbf{m}x + \mathbf{b}$, $y = \mathbf{a}x^2 + \mathbf{b}x + \mathbf{c}$, etc. — affected their graphs. We also wanted students to understand that different algebraic forms may make it easier or harder to see how a function's graph would behave — for example, that parabolas can be written in the form above, or as $y = \mathbf{a}(x - \mathbf{b})^2 + \mathbf{c}$, or (when the parabola has real roots) as $y = \mathbf{a}(x - \mathbf{b})(x - \mathbf{c})$, and the different forms make different properties of the graphs easier to see. For the former goal, we have a real–time link between the algebraic representation of a function and its graph. Consider a straight line, for example. When the student calls up the form $y = \mathbf{m}x + \mathbf{b}$, it appears both algebraically and graphically on the screen. (The default values 1 for **m** and 0 for **b**; the graph of $y = x$ appears on the screen.) The student can either type in values for **m** and **b** or use a "slider" (much like the slider volume controls on many stereo systems) to vary the values of either parameter continuously. For example, the student can type in 3 for **m** and 4 for **b**, at which point $y = 3x + 4$

appears on the screen. Then the student can vary **m** continuously — and as she does, she sees the straight line $y = \mathbf{m}x + 4$ rotating around its y–intercept, the point (0,4). If she fixes **m** (say **m** = 2) and varies **b**, she sees a variable line, with slope 2, rising or falling through the point (0,**b**) for whatever value of **b** she has on the slider.

Things get more interesting for the quadratics. We have the three algebraic forms for vertical parabolas [$y = \mathbf{a}x^2 + \mathbf{b}x + \mathbf{c}$, $y = \mathbf{a}(x - \mathbf{b})^2 + \mathbf{c}$, and $y = \mathbf{a}(x - \mathbf{b})(x - \mathbf{c})$] all displayed and all linked dynamically, to each other and to the graph. As the student varies one form, the others change simultaneously. Using the standard form $y = \mathbf{a}x^2 + \mathbf{b}x + \mathbf{c}$, the student can see that changing **c** results in a vertical translation of the parabola — but the effects of varying **a** and **b** are not so easy to understand. With the second form, one sees clear effects (changing the width, translating horizontally, translating vertically) for changes in **a**, **b**, and **c**, respectively. Similarly, the third form gives clear information about the roots. But the linkages also work between algebraic forms. When you operate on one of the first two and you move the parabola above or below the x–axis, the third form displayed changes to $y = \mathbf{a}(x - \mathbf{***})(x - \mathbf{***})$, indicating that you can't factor the equation of a parabola that has no real roots. With these and other dynamic, real–time tools, we hope to help students see some of the connections that are so hard for them to see otherwise.

Some non–standard consequences of learning to program.

For purposes of this article, programming *per se* (independent of whatever virtues there might be in learning to program) is not of interest. Some non–standard consequences of learning to program may be, however.

As we shall see in the discussion of microworlds below, Seymour Papert argues that programming environments may make certain concepts "come to life" for students. A child programming in Logo who wants to draw houses or flowers of different sizes, may draw them all from scratch — or the child may learn to define the procedures in terms of variables, which makes life a lot easier. For example, one can define a square box in Logo–like pseudo–code:

```
Define BOX(SIDE) as follows:
    Repeat the procedure
        [Draw a line SIDE units long and turn right 90
        degrees]
    four times.
```

Having done so, one can draw a BOX by specifying the value of the variable SIDE: BOX(3) produces a square three units on a side, for example. One can also concatenate commands: HOUSE(SIDE) can be defined as "Put TRIANGLE(SIDE) on top of BOX(SIDE)," where TRIANGLE is defined as an equilateral triangle. And if you want to draw pairs of figures in specific ratios, you can define BIGHOUSE(SIDE) as HOUSE(2*SIDE). Papert argues that in such ways students come to understand the idea of variable.

Ed Dubinsky [7] argues that programming experience can yield similar results with college students. He gives two cases in point. The first deals with composition of functions. Using the Unix operating system, Dubinsky had students engage in a number of exercises using shell scripts and pipelines. A *shell script* is a means of storing operating system commands in a file with a single name, so that entering the name at the terminal results in executing the entire sequence of commands. If the shell script for $h(x)$ — again in pseudo–code — is "Do $f(x)$, then do $g(x)$ to what you get," then $h(x)$ is in essence $g(f(x))$. A *pipeline* $g|f$ is a command that links two commands f and g, with the output of f serving as the input of g. Here too, if f and g are functions, $g|f$ is $g(f(x))$. Dubinsky claims that by virtue of their computer experience with these notions — trying them out as commands, and seeing what they produced — his students developed much better understandings of function composition than students do in standard instruction.

He also claims that experience with a computer language called SETL [6] helps students to understand quantification in first–order predicate logic. SETL is a high level procedural language that supports (among other things) universal and existential quantification over finite sets. Thus a statement (once more in pseudo–code)

Define A as:
$A = \{ (x,x^2)$: x in $\{1,2,3,...100\}$ and for all y in $\{2,3,...,x–1\}$, $x \neq 0 \bmod y \}$

serves to define A as the set of ordered pairs (x,x^2) with the property that x lies between 1 and 100, and x is not divisible by any of the integers less than x. That is, A is the function that assigns to each prime less than 100 its square. Hence $A(11)$ yields 121, but $A(15)$ is undefined. Figuring this out is non–trivial. Dubinsky argues that figuring out how SETL evaluates $A(x)$ calls for understanding the symbolic notation — and that students learn how to do so quite well in the computing context.

In general there is the question of whether, by creating mathematical worlds on computers that reflect basic mathematical ideas, we can give students "concrete" experiences that help them to master those ideas.

Intelligent tutoring systems.

The classic reference on this topic is Sleeman and Brown [15]. Because of space limitations I can only outline a few main ideas and point to a few examples.

In brief, the "best of all possible worlds" scenario for intelligent tutoring systems is as follows. Imagine building a computer environment sufficiently knowledgeable and flexible that the system could serve as a private tutor for a student. The system would present information to the student; the student would work practice problems; the system could speed the student along when her work was going well, but could also diagnose the student's mistakes and help when things went wrong; and it could answer the student's questions on a wide range of related issues. In order for any tutor (machine or human) to succeed at this, it has to be (a) expert at the subject matter, (b) pretty good at figuring out what's going on in the student's head, and (c) pretty good at teaching (i.e., have accessible a wide range of teaching strategies).

Needless to say, at this stage in their development computer–based tutors hardly approximate the "best of all possible worlds" status.[1] Some slight progress has been made, however. At the level of expertise — making sure the tutor knows the subject matter — the field of artificial intelligence has produced "expert systems" that rival human performance in such complex domains as medical diagnosis (for meningitis–related diseases) and the evaluation of NMR (nuclear magnetic resonance) spectra. At the level of diagnosis, there has been less, but still some notable, success. Perhaps the best known is Burton and Brown's work [4] on students' "bugs" in elementary arithmetic. There are various causes for students' mistakes in addition and subtraction problems, one of which is simply that the student hasn't yet fully understood the procedure. But another is that the student may have misconstrued the procedure and is systematically doing the same thing over and over again, getting the wrong answers not by inadvertence, but by having "mastered" the wrong thing. Burton and Brown studied students' work in detail and found a number of consistent error patterns. With a computer–based analysis of a sixteen item diagnostic test, they are now able to predict, about half of the time, not only whether a student will get new problems right or wrong, but the precise answers that the student will give to the problems. They make their predictions before the student works the problems, of course. The results indicate that (a) the students are doing things pretty systematically, even when they're wrong, and (b) Burton and Brown have a pretty good model of what's going on in the students' heads when they work the problems. At the level of pedagogy, I have to confess that there are some ad hoc principles for good tutors, but nothing rigorous or well defined. Artificial Intelligence is still in its youth, and intelligent tutoring systems are in their infancy. (Sleeman and Brown [15] is really the first compilation of articles on the subject.)

Even so, there are some signs of potential success, such as tutorial systems for symbolic integration, electronic troubleshooting, the interpretation of NMR spectra, medical diagnosis, solving algebraic equations, gaming environments, program–plan debugging, writing proofs in Euclidean geometry, and learning to program LISP. All of these are prototype systems in rudimentary phases of development. Yet there is evidence that some of them serve as pretty good learning environments for students. Two such environments are as follows.

WEST ("How the west was won;" see [3]) is a board game like "chutes and ladders," where one tries to advance along a road 70 squares long, avoiding pitfalls (which cause you to move backwards) and taking advantage of bonuses (squares that move you ahead). When it's your turn, three small integers A, B, and C turn up on three spinners. You can move ahead any arithmetic combination of A, B, and C — as long as you specify it. So if you get (2,1,2) on the spinners, you may move ahead (1+2+2), (1+2–2), (1+2×2), (2×(1+2)), etc. — whichever is best, given the conditions of the game.

As students play the game, the tutor sees their "moves" — their arithmetic combinations of the three numbers that turned up on the spinners. The tutor then has to decide (a) whether the student is using the arithmetic operations to get the results he or she wants, and (b) whether the student's strategy could be improved. And the tutor has to decide when to intervene and

---

[1]The typical live alternative hardly represents the best of all possible worlds, either. The school alternative often consists of one teacher trying to manage a classroom of thirty children all at once — how much individual attention does one child get? And many college lecture classes have hundreds of students getting little or no individual attention.

suggest that the student go over something or try something new. Burton and Brown ran an informal study with a small number of student teachers playing the game. "Eight of ten subjects [who had the computer coach comment on their work] found the comments helpful in learning a better way to play the game, and nine out of ten felt that *the Coach manifested a good understanding of their weaknesses!*" [3, p.97]. In classroom studies, students who used the WEST tutor showed the optimal strategies more frequently and more reliably than students who played the game without the computer–based tutor.

A second tutor is radically different, both in terms of the subject domain and in tutorial style. John Anderson [2] has developed a tutor to help students write formal proofs in Euclidean geometry. It is based on the insight that the standard two–column proof format presents proofs as being linear, which obscures the non–linearity of the search process. That is, in solving a proof problem, one usually checks to see what one can get from the givens, looks to see what will produce the desired result, and then tries to narrow the distance between the two. In formal terms, this dialectic ultimately results in "finding a path through the search space." Only when you have the path can you traverse it linearly. Anderson's tutor contains an "expert" that solves the problems first and then constrains the student to follow "ideal" solution paths, teaching the student (in theory) to solve the problems like the expert does. In preliminary studies with the tutor, three students who had failed a standard Euclidean geometry class spent a summer working with the tutor and passed a geometry exam with flying colors. In similar studies with Anderson's LISP tutor, students who studied particular units using the tutor (including learning to program recursion) learned faster and did better on standard exams than students taking a lecture course at Carnegie–Mellon University.

I shouldn't paint too rosy a picture, for there is plenty of reason to be unhappy with all computer–based tutors that exist today. In particular, I am very uncomfortable about Anderson's tutors: They get students to master certain procedures in geometry, but I'm not at all sure how much the students understand. But if the choice is between having students who understood little and couldn't write proofs, and students who understand little and have mastered proofs ... well, I'm reminded of the story of the chess–playing dog that won its first game, after playing ten with a local expert. "The wonder of the thing isn't that the dog finally won a game. The wonder is that it could play at all." By any reasonable standard, intelligent tutoring systems have a *very* long way to go. But by any reasonable standard, we have made astonishing progress to get where we are.

Computer–based microworlds.

Papert [12, p. 6] introduces the concept of microworld as follows:

> It is possible to design computers so that learning to communicate with them can be a natural process, more like learning French by living in France than trying to learn it through the unnatural process of American foreign–language instruction in classrooms ... We are learning to make [mathematics–speaking] computers with which children love to communicate. When this communication occurs, children learn mathematics as a living language. ... The idea of "talking mathematics" to a computer can be generalized to a view of learning mathematics in "Mathland"; that is to say, in a context which is to learning mathematics what living in France is to learning French.

Papert's creation, Logo, is the best–known microworld. Here we focus on the "Turtle Geometry" part of Logo, in which a "turtle" can be programmed to move around a computer screen. The turtle leaves a trace of its movement when you command it to; hence you can draw pictures by programming the turtle's movement. Logo is a LISP–like language with some very simple primitives, such as moving forward or back a specified number of units, or rotating left or right a specified number of degrees. But the language also includes variables, function calls, recursive definition, and other powerful constructs.

At the upper end of the spectrum, Logo is a rich computational universe allowing for some high–powered and interesting mathematics. In Abelson and diSessa [1], for example, one finds: recursive programs generating Hilbert's and Sierpinski's space–filling curves; topological deformations of two– and three–space; and a proof (using the "intrinsic geometry" of turtle mathematics, rather than the "extrinsic geometry" of standard coordinate systems) of the Gauss–Bonnet theorem linking the total curvature and the Euler characteristic of any closed surface.

At the lower end of the spectrum, young children can draw pictures of houses and trees by programming the turtle. They can start on a line–by–line basis, writing "spaghetti code" — linear descriptions of how to trace out the pictures they want. But they soon discover that long descriptions contain lots of mistakes; it's easier (and better) to break things into subprocedures. (For example, you can draw a house by making a box and putting a triangle on top of it.) Later students learn about variables: If they want to make houses of different sizes, they can make the boxes and triangles of side X instead of a fixed size. In short, Logo is an "easy entry" world that contains mathematical riches. The idea behind its design is that when students learn how to live in that world, they learn important mathematical ideas: the notion of variable, of procedure, etc. They also learn some important meta– mathematical ideas: that answers can be "on the right track" and can be de–bugged (as opposed to the "it's right or it's wrong" view they get in the world of arithmetic), that doing things in systematic and modular form can be useful, etc.

Logo is hardly a panacea, and the marvelous things suggested in the previous paragraph may or may not happen with very many children. What's important here is the concept, that of a mathematically rich microworld. We now have an existence proof that such things are possible. The papers by Jones and King in this volume present the two rather different college–level courses based on Logo.

Transcendent technologies and general all–purpose mathematical tools.

This last example is about as far away from mathematics as you can get, but it illustrates an important point. The analogy to be exploited here is that technologies come in two varieties: those that allow us to do things we could do before, but faster, better, or more efficiently; and those that enable us to do things we simply couldn't do before — the transcendent technologies. While we normally think of technologies as physical tools, there are cognitive technologies as well. Far and away the most important cognitive technological revolution was the invention of writing and the use of written records. Try as you might, you're not going to be able to multiply 92736 times 47863 in your head; the task calls for keeping far more things in mind than your short–term memory buffer can hold (without "chunking," which will make the task possible if you practice enough). Moreover, what you "know" in a world without written records is limited to what you've committed to memory or can derive on the spot. Thanks to writing, of course, Gauss can be our private tutor. Written records, a cognitive technology, allow us do to things hitherto impossible.

In our professional lives, computer–based technologies have begun to have such transcendent (although not uncontroversial!) effects. There are, as yet, no transcendent mathematical tools for societal use in general. In fact, there are precious few examples at all. One, however, is worth noting as a prototype. Text processing — relatively cheap, and about as easy as typing on some of the more user–friendly systems — has effected significant changes in the ways some people write. Of course, text processing allows for more efficient editing; with virtually no effort, I've edited this paper on–line as I wrote it, moving sections of the paper around as I decided to change the order of presentation. At that level, the text processor makes it (much!) easier for me to do what I'd do anyway. But for others, text processors help to overcome some writing difficulties. For example, a technique called "free writing" consists of generating text uncritically and then editing — the writing equivalent of "brainstorming." There is some evidence that people who can't do free writing with pencil and paper (because what they're writing stares them in the face) can do it with a keyboard. This technique allows them to overcome writer's block. Some systems include "outliners" that make it easy to outline papers. Much more interesting systems currently under development (for example, "Note Cards" at the Xerox Palo Alto Research Center) are, in essence, writers' databases — "idea management" systems that allow the user to keep track of ideas, commentary, information, etc., with ease. None of these tools will ever make writing easy, but they are making it easier and will continue to do so. Moreover, such tools are not the exclusive property of professionals. They are becoming increasingly accessible to the general population.

Since there are no comparable examples for mathematics, the issue here is a visionary one. Can we develop systems for helping people to think mathematically? Imagine a mathematical tool kit, as flexible and useful for helping people to think mathematically as the text processor and related tools are for helping people to write. Pie in the sky? Maybe so, but remember this: In 1977, Ken Olson, the president of Digital Equipment Corporation, made the following statement at the Boston convention of the World Future Society: "There is no reason for any individual to have a computer in their home."

A brief postscript.

As I noted at the start, we mathematicians are but slowly becoming acclimated to the use of computers in our professional lives. We have been even slower to use technology in the classroom, and our use has for the most part been unimaginative. [This is no surprise. Recall that the first "horseless carriages" looked just like the carriages that horses pulled, simply because carriages were people's models of "moving vehicles." It took a while for cars to evolve their own shapes.] I believe it is possible to harness computer–based technologies in interesting ways, ways that go far beyond allowing us to do thing faster and cheaper than before. With innovative use of technologies, we may be able to do things that, until now, were simply not conceivable. I hope the examples in this paper point to some interesting directions for such work.

Acknowledgements.

The work described in this paper was supported by the National Science Foundation through NSF Grant MDRP8550332. NSF grant support does not necessarily imply NSF endorsement of the ideas expressed in this paper.

REFERENCES

1. Abelson, Harold, and Andrea diSessa. **Turtle Geometry.** Cambridge: MIT Press, 1980.

2. Anderson, J., C. Boyle, F. Franklin, and B. Reiser. "Intelligent Tutoring Systems." **Science 228** (1985), 456–462.

3. Burton, Richard, and John Brown. "An investigation of computer coaching for informal learning activities." In Derek Sleeman and John S. Brown (Eds.), **Intelligent Tutoring Systems**. London: Academic Press, 1982.

4. Burton, Richard, and John Brown. "Diagnostic models for procedural bugs in basic mathematical skills." **Cognitive Science 2** (1978), 155–192.

5. Brown, John, Richard Burton, and Johann deKleer. "Pedagogical, natural language, and knowledge engineering techniques in SOPHIE I, II, and III." In Derek Sleeman and John S. Brown (Eds.), **Intelligent Tutoring Systems**. London: Academic Press, 1982.

6. Dewar, R., E. Dubinsky, E. Schonberg, and J. Schwartz. **Higher Level Programming**. New York: Springer. (In press.)

7. Dubinsky, Ed. "A new approach to teaching abstract mathematical concepts." Manuscript submitted for publication, 1986. Available from author, Department of Mathematics, Clarkson College.

8. Dugdale, Sharon. "Green Globs: A microcomputer application for graphing of equations." **Mathematics Teacher 75** (3) (1982), 208–214.

9. Dugdale, Sharon. "Computers: Applications unlimited." In V.P. Hansen (Ed.), **Computers in Mathematics Education** (1984 Yearbook of the National Council of Teachers of Mathematics), pp. 82–88. Reston, VA: National Council of Teachers of Mathematics, 1984.

10. Hollan, James D., Edwin L. Hutchins, and Louis Weitzman. "STEAMER: An interactive inspectable simulation–based training system." The AI **Magazine** (Summer 1984), 15–27.

11. Nickerson, Raymond S. (Ed.). **Technology in Education in 2020: Thinking About the Not–Distant Future**. Hillsdale, NJ: Lawrence Erlbaum Inc. (In press.)

12. Papert, Seymour. **Mindstorms: Children, Computers, and Powerful Ideas**. New York: Basic Books, 1980.

13. Pea, Roy D. "Cognitive technologies for mathematics education." In A. Schoenfeld (Ed.), **Cognitive Science and Mathematics Education**. Hillsdale, NJ: Erlbaum. (In press.)

14. Schoenfeld, Alan H. "Mathematics, technology, and higher–order thinking skills in the near and not–so–near future." In Raymond S. Nickerson (Ed.), **Technology in Education in 2020: Thinking About the Not–Distant Future**. Hillsdale, NJ: Lawrence Erlbaum Inc. (In press.)

15. Sleeman, Derek, and John S. Brown (Eds.). **Intelligent Tutoring Systems**. London: Academic Press, 1982.

16. Swift, Jim. "Exploring data with a microcomputer." In V.P. Hansen (Ed.), **Computers in**

Mathematics Education (1984 Yearbook of the National Council of Teachers of Mathematics), pp. 107–117. Reston, VA: National Council of Teachers of Mathematics, 1984.

17. Yerushalmy, Michal, and Richard Houde. "The geometric supposer: Promoting thinking and learning." **Mathematics Teacher 79** (6) (1986), 418–422.

RESOURCES

Some of the software discussed in this article is commercially available. For information about the symbol manipulation systems MACSYMA, Maple, muMath, Reduce, and SMP, see John Hosack's "A guide to computer algebra systems" in the November 1986 **College Mathematics Journal**. More detail about any of the systems discussed here can be obtained by writing the authors cited in the references. In addition to the algebra systems reviewed by Hosack, four of the software items mentioned in this article are available.

The Geometric Supposers can be purchased from Sunburst Communications. Information about the Supposers and other ETC software can be obtained from the Educational Technology Center, Harvard Graduate School of Education, 337 Gutman Library, Appian Way, Cambridge MA 02138.

Logo is available in many versions, for many machines. One general source of information about Logo is **Logo Exchange**, which is published monthly September through May by Meckler Publishing Corporation, 11 Ferry Lane West, Westport CT 06880.

Information about Bob Tinker's software and other TERC products can be obtained from Technical Education Research Centers, 1696 Massachusetts Avenue, Cambridge, MA 02138.

Information about Green Globs and related products can be obtained from Sharon Dugdale, 252 Engineering Research Lab, University of Illinois, 103 South Mathews, Urbana, Ill., 61801.

Evaluating Mathematical Software

R. Stephen Cunningham

This article is a guide for mathematicians who want to start using the computer in their teaching or to use it more widely. I assume the reader already understands a computer's capacities and believes the computer can be valuable in teaching. My goal is to help the reader select the best software to use with his or her teaching.

The ideas in this article represent far more than my experience as an author, referee, and editor in the area of evaluating software. It includes contributions from Judy Brown (The University of Iowa), Shelly Gordon (Suffolk County Community College), and Stu Thomas (Wadsworth Publishing Co.), participants in a panel discussion on evaluating mathematical software at the MAA meeting in New Orleans in January, 1986. Furthermore, it represents the collective experience of the many persons who have written software reviews for *The College Mathematics Journal*, many of whose comments are explicitly cited. I describe the need for careful software evaluation, give some criteria which can be used in evaluation, and discuss some machine and management issues which affect software use.

The need for software evaluation.

In the last few years the amount of software available to mathematics instructors has grown dramatically. A recently published software database [1] shows both a large number of college–level packages and a relatively wide curriculum coverage at the lower division, plus some coverage of the upper division. We are no longer limited to two or three pieces of software for a course. With a wide range of commercial and noncommercial products comes a considerable range in quality and usability. The main questions you must consider about software are little different from those used in evaluating textbooks: Does it cover your material (in full or in part)? Can your students learn from it? Can you teach from it? Is the material presented well?

While the amount of available software is large, it is quite possible that you may search carefully and still find nothing suitable. In this case you may want to write your own software. Your careful evaluation will still be of great help to you, as you will know very well just what you want from the software. Be aware, though, that writing good instructional software is a serious and demanding task.

Some evaluation criteria.

The best source of criteria for evaluating software is probably a collection of such evaluations by actual users. We are fortunate to have available a number of very fine evaluations in the software reviews published (or soon to be published) in *The College Mathematics Journal*. On examination of the reviews done over three years, five criteria stand out:

- usability in the course;
- usability by the student (user interaction and human factors);
- soundness of pedagogy used in the programs;
- quality and usability of the documentation;
- correctness.

We will expand on each of these below.

Course Usability

Most software is intended for use in a single course. A given package might cover a single topic or the entire course, and it might be very detailed or quite general. Here is a good statement of what makes software usable in a course [3]: "[S]oftware...must meet three tests. First, it must help students learn material that is difficult to absorb by traditional methods of learning. Second, it must help students develop a deeper understanding of the standard course material. And third, it must enable the instructor to introduce topics not usually contained in courses but which ought to be covered because of their importance."

Most instructional software falls into at least one of the following categories: drill and practice, simulation, or exploration. The instructor needs to plan the appropriate computing modes for the particular course and then determine whether the software being considered can operate effectively in those modes. Here is an example of the kind of feature that should be observed [4]: "When one display is produced on the screen the previous graphics screen is automatically saved. The user can toggle between screens by a single keystroke, making some beautiful demonstrations possible." Individual style is important in determining whether a given software product can be used in a course. We all recognize the wide range of course styles and the large number of products needed to meet these styles. (Consider the number of textbooks available for any one course.) You should be prepared to look at many pieces of software in order to fit your particular style needs.

Courses also vary in the level of presentation desired by the instructor to meet the needs of a particular class. Software shares this variation: It can present very general or quite detailed explanations, and it can require information at a high level or need a great deal of detail. You should try to match the level of the presentation to the level you want to use in your class.

Like textbooks, no software ever seems exactly right for a course (unless you write it yourself), so you should consider whether the software can be modified or adjusted to fit your needs. There are two ways this can be done: The software can have special setups for demonstrations or adjustment to the user's level which can be saved for future use. And the programs' source code may be available so you can change particular explanations or functions to match your needs.

Finally, a word about software distributed with a textbook. This mode of distribution is becoming more common, but such software is not always the best choice even with its accompanying text. This kind of software often is developed with less care than is given to software developed for its own sake. Software adoption must be based on the course and the instructor's needs, regardless of the text used.

Human Factors and Usability by Students

This area has received considerable attention in software reviews because of the uneven characteristics of available user interfaces and the increasing consciousness of users about the way software is used. The point was first emphasized by CONDUIT more than ten years ago, and it has been accepted more recently by other publishers. Quality in this area is certainly improving, but it cannot be taken for granted. In themselves, most human–factors principles are less important that the presentation of the subject matter. However, the way software meets or does not meet these principles can affect students' interest, understanding, and learning.

The most important principle is effective interaction with the student. The goals here are to keep the programs' instructions and results understandable and to allow the student's input to be natural. A couple of reviewers' comments in this area [5], [6]: "It is so easy to use and extremely well documented that students will enjoy their experience at the computer instead of feeling that they are doing battle with the machine" "The programs print fractions as fractions, corresponding to the usual approach to matrix operations." The programs should respond to input in any correct format and allow experienced users to move more quickly. As observed in [7], "Shortcut forms of the commands and the ability to turn off prompts and warnings allow the experienced user to speed program operation." The pace should be appropriate to the mode of use and controllable by the user; this helps maintain student interest and concentration. Slowing down unduly (say, on graphics) or seeming to stop during long computations with no indication of activity is undesirable.

Graphics aid the student's ability to learn from software, but they should be used carefully. From another review [8]: "Graphics is used where it can be effective, and is not used when it adds nothing to the presentation." We have an appreciation of good graphics from our own use of blackboard diagrams, and the computer can go well beyond this to show information clearly, correctly, and dynamically, but it takes real skill to design good displays. Factors involved here include good layout, clear presentation and labeling, and (sometimes) good animation. Color is often an important aid to the student's understanding, but this has to be distinguished from "showing off" with flashy color displays when monochrome would do just as well.

If programs are to be used with student– or instructor–provided data or functions, the ability to save this information for future use is critical. It is also very helpful if such information can be edited. Another reviewer's comment [9]: "Students can enter data from the keyboard with the option of storing the data on disk. Students are also allowed to edit existing data sets. These features can save students time on entering data when they want to run the programs repeatedly."

In spite of the best instructor's help and written documentation, errors will be made when students run programs. Help should be available — ideally, *always* available — and it should be really helpful, instead of just repeating text screens. From [10]: "A 'HELP' option is available while performing reduction steps...asking for help results in a statement telling what the next operation should accomplish, which operation will do it, and the correct row numbers and scalars to use." When errors do occur, the software should catch them without crashing and should do as much diagnosis of the errors as possible. Such a system is described in

[11]: "If an answer is not in acceptable form, a prompt will be given and the student given a chance to rewrite the answer. Syntax errors are caught and pointed out."

Finally, screen output is simply not enough for most educational software. Sometimes hardcopy is necessary to be handed in or to retain a record of a particular experiment. This can be text–only, or it can include graphic screen copies or plotter output. Hardcopy output should be considered in looking at software, and compatibility with your printer is important.

Program Pedagogy

One important pedagogical principle is that the level of any presentation should be adaptable to the student's needs; in particular, this is useful for software. As observed in [12], "Various features...can be changed to make [the program] easier or more difficult."

Good software will also show us new ways to approach the teaching of familiar topics, which can be significant additions to the course and to students' understanding. Examples from reviews [13], [14]: "One experiment comes to mind immediately...(it) allows one to zoom in on the graph of a function near a given point and see the fundamental difference between a function which is differentiable at a point and one which is not. In this mode, the program allows the student to explore his or her own ideas." "Assignment 5 is a complete series of questions involving the verification of trigonometric identitites using computer graphics. This encourages the development of problem–solving skills in an innovative and fun way."

One of the most promising kinds of software allows the student to explore new ideas in an unstructured way. Such true exploration tools are still unusual, but the use of such a program is described in [15]: "[I]t emphasizes the student's freedom to discover new ideas... The student becomes an active participant in the discovery of geometric notions, rather than simply a passive observer of pre–established facts."

Finally, as you consider using software in support of a mathematics course, look for tools that let students work flexibly and do not force rigid work operations.

Program Documentation

Few programs can be run successfully with no information on their use, and few individuals can run a program successfully without instructions. "No worthwhile package is only a collection of programs. It must have a manual which can support and guide the user through the programs." [8]

Documentation can be on paper in an accompanying manual, or it can be presented on–line, in both instruction screens and help materials. This documentation must contain complete, accurate, and clear instructions on the program itself. Beyond this, however, it should describe the mathematics it supports and the underlying principles of that mathematics. For example [16]: "The documentation is detailed, friendly, and complete... [it] describes the specific contents of this part... [and] guides the new user through a step–by–step tutorial."

On–line help should be seen both as part of the documentation and as a user interface tool. Such help is the best possible source for immediate information, and it should be considered carefully in evaluations.

Correctness

There are two aspects to the question of correctness: the accuracy of the information given to the student and the soundness of the programs' computation. If either of these fails, the software is clearly flawed. Incorrect information is the lesser of these two problems. The instructor can make up for these errors by providing corrective information, but this is very distracting and requires extra work from the instructor.

Computational errors, though unusual, can happen and must be avoided. They can come from poor theory, such as unsound statistical approaches, from poorly designed algorithms, or from poorly controlled roundoff errors. Roundoff errors, in fact, are a continual problem in software; any program doing heavy calculations should have routines that minimize their effect, and its documentation should describe what such errors might be.

Machine support.

In evaluating software you must consider the computer environment in which it will be used. If you already have equipment available and have no plans for a change, then you should only consider software for that equipment. Our research for the software database [1] showed that there is very little college–level instructional software available for machines other than the Apple II family, IBM PC's and compatibles, and the Macintosh, although enough is now appearing for the

Commodore 64 and 128 and the TRS–80 to warrant their inclusion in the second installment of [1]. Beyond this, the following observations on the future directions of hardware and software should be considered.

First, at some time in the near future it may become more effective to use video projectors rather than large–screen monitors for demonstrations. The cost of projectors is coming down, their capabilities are improving, and the better ones allow the instructor to maintain better control of the demonstration. Some video projectors are not compatible with all computers, however, especially at the low end of the cost range, so be careful to check this before purchase. Even if the projector is compatible, its image may not be sharp enough to display 80–column text or bright enough to use in a normally–lighted classroom. Most only display monochrome images and some color combinations lose their contrast. Really satisfactory projectors are rarely in the price range which is reasonable for educational institutions.

Second, there is a new class of potentially exciting classroom display devices: liquid crystal screens to be placed on overhead projectors. They are available for the IBM PC with Color Graphics Adapter or equivalent and for the Apple // with RGB color. They list for $900–$1200 and are very portable. They have the further advantage that the instructor can write on a blank transparency sheet placed over the device to add comments to the computer display. However, these devices may not work well with some color combinations, since software developers normally assume their images will be displayed on a monitor, not on a color translation device. You may have some surprises unless you can control the colors by a setup file or by switching to monochrome display mode. Since they use an overhead projector, these devices require the same lighting adaptation that such projectors usually need.

Third, newer software favors the IBM PC/MS–DOS family over the Apple II family. However, instructional machines need graphics, especially color, and some retrofit may be necessary if an existing lab is to be used. A new lab right now should probably use IBM PC's or compatibles, but this recommendation must be continually reviewed. The world has a way of making such statements obsolete, and other machines, such as the Macintosh (and maybe the Amiga and/or Atari ST series) with their superior speed and graphics, could become more important.

Management issues.

The major issues of management are the interlocking questions or cost, software copying, and copy protection. The problems here are straightforward:

- some software *is* too expensive;
- using multiple copies of a single purchased piece of copyrighted software *is* illegal; and
- copy protection *does* make it awkward to make legal backup copies or use software with hard disks.

These issues are discussed very effectively in [2].

Software publishers are generally responding to demand by lowering their prices to levels more comparable to textbooks. Good software is even being developed by a few individuals who share it with others at their distribution cost. With appropriate pricing, departments should purchase the appropriate number of packages to support their course or have each students purchase a copy. This would help break the cycle of publisher protection and instructor copying that has unfortunately been so prevalent in academia.

Some publishers have made site licensing the standard sale condition for educational software, a policy we are pleased to see. It is not unreasonable to find a higher price for such software than you would expect for a single copy; this benefits both the institution and the publisher who incurs a significant cost in developing really good materials. If the published price for the number of copies you want is unacceptable, you should always negotiate for a bulk rate or a site license. Most publishers are willing to make a reasonable deal. This statement also applies to licensing for a campus–wide network or a local area network (LAN). You may not find any published network price for the software you want, but the publisher would rather negotiate on the basis of the expected use than lose the sale entirely.

In any case, we must regain the respect for the copyright laws that was too easily forgotten in the days when computers were so expensive that software was viewed as free. By cooperating with authors and publishers we should be able to arrange pricing and licensing policies that reflect both the genuine cost of developing good software and the limited budgets of departments.

A management issue separate from copyright, but involved in it because of copy protection, is the use of hard disks. Individual machines' hard disks seem to accumulate student files and tie the student to a particular machine, a problem if that machine is busy or goes down. A better use of a hard disk in a laboratory setting is to be the central file server in a LAN. This allows student files to be monitored, and you can develop your own policies for student files on diskettes or the central system. However, most purely instructional systems work quite well on diskettes; it's not clear that hard disks are really needed in a laboratory. On the other hand, a hard disk is extremely useful in a computer installed in a classroom for use by the instructor. All the appropriate software for all the courses using that room can be made available at the touch of a few keys, and awkward delays that result from changing programs or accessing files on floppy disks are eliminated. A menu system for accessing the contents of the hard disk makes it easy even for instructors who are unfamiliar with computers to do effective classroom displays.

REFERENCES

Articles:

1. Cunningham, R. S., and David A. Smith. "A Mathematics Software Database." *The College Mathematics Journal* **17** (1986), 255–266. "A Mathematics Software Database Update." *The College Mathematics Journal* **18** (1987), 242–247.

2. Brooks, Daniel T. "Copyright and the Educational Uses of Computer Software." *EDUCOM Bulletin*, Summer 1985, 6–13.

Software Reviews in *The College Mathematics Journal*:

3. Herman, Eugene A., review of *Compucalc*, forthcoming.

4. Matchett, Andrew, review of *Probability and Statistics Demonstrations and Tutorials*, forthcoming.

5. Leinbach, L. Carl, review of *ARBPLOT*, **15** (1984), 160–162.

6. Lamet, Daniel G., review of *Computer Applications for Finite Mathematics and Calculus*, forthcoming.

7. Hannula, Thomas, review of *Matrix Algebra Calculator*, forthcoming.

8. Leinbach, L. Carl, review of *Calculus Toolkit*, **17** (1986), 90.

9. Chang, Stephen, review of *Introductory Statistics Software Package*, forthcoming.

10. Edwards, Sheryl, review of *Matrix Reducer*, **18** (1987), 159.

11. Rossi, Don, review of *Algebra Drill and Practice I and II*, **16** (1985), 222.

12. Hirschfelder, Rosemary, review of *Algebra Arcade*, **18** (1987), 158–159.

13. Abram, Thomas J., review of *Interactive Experiments in Calculus*, forthcoming.

14. Schmalzried, Cynthia, review of *Discovery Learning in Trigonometry*, **15** (1984), 263–264.

15. Kunkle, Dan, and L. Carl Leinbach, review of *The Geometric Supposer*, **18** (1987), 66.

16. Brown, Judith R., review of *WEPCO Electronic Blackboard Series*, **17** (1986), 358–359.

The Machine in the Garden: Calculus with Computing

L. Carl Leinbach

There once was a beautiful garden with sweet smelling flowers, lovely trees, and lush bushes. People enjoyed visiting the garden and walking its paths, where they found beauty and solace. One day a machine appeared in the center of the garden. The machine produced vials of perfumes that smelled as sweetly as the flowers and pictures of scenes which rivaled those in the garden. People took the vials and the pictures. They no longer returned to the garden. It soon became overrun with weeds because there was no one to care. The only memory of the beautiful garden that remained was the one generated by the machine.

Is this little story a metaphor for the relationship between the calculus (a beautiful intellectual garden) and computing (which uses a powerful intellectual machine)? In this essay I present a view that computing can be used to amplify the results of the calculus and to help students discover and test results of the calculus in much the same way that a physics or chemistry student uses the laboratory to discover and test scientific laws. Computing is not the substitute for doing calculus. It is a tool which is available to enhance the subject. Although there is a danger of misusing the tool, my premise is that computing can help us to teach a better, more lively calculus course.

The use of the computer as a tool in the teaching of calculus is one with a long established tradition. Papers on the use of computing in calculus courses started appearing in the mid–1960's, and there were privately and federally funded projects to aid teachers in implementing the use of computers in the late 1960's and into the 1970's. These projects included the establishment of computer laboratories, development of course materials, and workshops for teachers of first and second year calculus courses. The supported projects produced volumes of materials and received attention at national meetings of the MAA and other organizations.

However, despite the long tradition, the activity, and the initial support, the use of computing in calculus has been, until recently, an uncommon activity. One reason is that this reflects the position of computing within the general mathematical community. In this view, mathematics involves reasoning within the framework of an abstract axiomatic system, while computing deals with particular results. Mathematics searches for universal statements, and computing produces algorithms of limited generality. Programs which mimic abstract reasoning are slow, difficult to write, and contribute little to the body of mathematical knowledge. As far as the use of the computer in the classroom to illustrate specific results was concerned, the prevailing opinion was that gains made in course content were offset by the effort required to instruct students and faculty in the techniques of computer use. The particular villains were programming languages and arcane operating systems.

The history of computing in calculus offers an excellent case study for the more general question of computing in any mathematics course. What is gained and what is the cost? If calculus is properly presented, it offers a unified development of a body of knowledge that has had a profound influence on the history of the human race. Its many and varied applications only add to the beauty of the subject and further stress its universality. It is argued that the introduction of computing into the course diverts attention from this development. Fundamental to this development is an understanding of the concept of a function defined on $\mathbf{R}^n$ and the meaning of the results of processes applied to these functions. Calculus deals with continuous processes; computing is by its very nature discrete.

The proponents of computing in calculus make the point that the purpose of their efforts is not to undermine the goal of the calculus course but to motivate many of the concepts and provide an aid for understanding them. By writing the algorithm contained within the development of a concept as a program, not only the concept but also its development will become clear. The discovery of an algorithm for a construction may lead to a more general result. Graphical output may lead to visual reinforcement, and the use of appropriate numerical techniques may lead to the solution of problems that are more realistic than those found in most calculus texts. Thus, the use of the computer can strengthen the course. Many of the recent developments in software have made the time needed for instruction in use of the computer minimal.

The debate continues, and I will not address its fundamental question. I assume that there is a technology that can be used to advantage in the teaching of calculus. Its use cannot be haphazard. It must be planned and implemented in a manner consistent with an individual's teaching style and the available facilities. Which applications are appropriate? Should students be required to write programs? Are graphical illustrations to be used in lecture? For what topics? If a symbolic manipulation program is to be used, which one? How heavily should it be emphasized? When should it be introduced? How much reliance should the student place on it? Answers to these questions will vary from instructor to instructor and from course to course.

I will survey the past work in the use of computing in calculus and emphasize those projects that may be appropriate in present courses. I will then discuss some of the software currently available for calculus instructors. In the final section, I discuss possibilities for incorporating computing in existing courses and some strategies for a redesigned calculus sequence.

Past Uses of Computing in Calculus.

Most of the early uses of computing in calculus were of a numerical nature. This was because graphical routines were not available on most machines, and those that were available were not standardized and thus not very portable. Symbolic manipulation programs were in their infancy and not widely known.

Common computing exercises asked students to generate approximations to limits, derivatives at a point, and definite integrals. Other problems called for the evaluation of partial sums of series and Taylor polynomials. In most cases the approximation techniques were crude, and the purpose of the exercise was to emphasize the definition of the object under consideration. Error estimates were used to control the output, but the efficiency of the algorithm was not a major concern. In fact, the use of too sophisticated an algorithm would defeat the purpose of the project. The goal was to have the student teach the algorithm to the computer and thereby come to an understanding of the concept.

This philosophy was carried one step further by Gerald Porter [6], [7]. He did not present the algorithm to the students. Instead, he asked his students to write programs to solve problems and then discussed the algorithms that the students developed. By this method more general mathematical constructions were developed. Example problems used by Porter are: find approximations to a number whose square is 2, locate a number between 2 and 3 such that $x^3 + x^2 - 5x + 5 = 0$, and approximate a number whose sine is 1. Another assignment asked the students to approximate the area under a curve. The goal of these assignments was not an accurate value efficiently calculated. It was the programs themselves! The student algorithms were converted into methods of constructing the real number system, proving the intermediate value theorem, and developing the definite integral. If a student developed a bad or overly specific algorithm, it was still possible to discuss the algorithm and analyze its weakness. The student was involved with the subject and was neither a passive observer nor overwhelmed by a theoretical argument concerning a topic that had not been confronted. The necessity to write a program forced the student to think about the topic. Porter's approach introduced algorithmic mathematics into the classroom in a natural and meaningful way.

While Porter's approach emphasized the development of the algorithm, an approach developed by a group at Duke University headed by David Smith developed software that involved the student with the use of the computer but eliminated the need to write programs. The software consists of a BASIC language shell in which the user types lines that define a problem in the form of sequences to be generated by the computer. It is used in conjunction with Smith's book [10], which describes projects and has exercises designed to guide the student's analysis of the computer output. The content of the standard course is not changed, and the computer laboratory serves the calculus course in much the same way the science laboratory serves the science course. The students perform experiments that illustrate the concepts taught in lecture and the text. By minimizing the need to program, the student is involved with the content of the calculus course and not the mechanics of program writing. Further details of this approach and its underlying philosophy are given in Smith's article elsewhere in this volume.

Applications have always played an important role in the first year calculus course. Computing has made it possible for the instructor to use more realistic and more interesting applications. By using elementary numerical techniques, it is possible to examine the growth of a population, watch the spread of a rumor, observe the interaction between a predator and its prey, analyze an arms race, examine the flow of traffic, determine the position of a submarine from reading a strip of graph paper passing under a stylus attached to a spring, find the area of an irregularly shaped lake, or estimate the average life expectancy for a class of individuals.

Projects may be as varied as the imagination of the instructor and are not confined by the need to avoid "messy" calculations. The student needs only to decide on an attack for the problem and analyze the result.

Interesting applications have the effect of stimulating interest in the calculus, and they lead to questions that in turn lead to the need for extensions of results. For example, what happens if one places a bounty on the predator population? How long will it take for a reduction in the birth rate to have an effect on population growth? What if the timing of a stop light were changed? While the introduction of these questions may have the effect of complicating the mathematical analysis, they may be tested by the inclusion of an "if" statement into the program. It is this property that John Kemeny refers to as the "what–if factor." Students can modify the model in an endless variety of ways and analyze the results with the same ease as they can analyze the original model. By encouraging students to satisfy their curiosity in one setting, it is natural to assume that the same curiosity will carry over to the subject that provided the tools for the construction of the model.

There was certainly no shortage of good ideas for the use of computing in calculus. Numerical approximation, the discovery of mathematical insights, a study of the effects of certain operations on functions, and the investigation of meaningful applications are all worthy activities and are, to some extent, found in most calculus courses. Yet, the idea of computing in calculus did not gain widespread acceptance. Inertia is one contributing factor, but not the only one. The calculus course is well established. The syllabus is generally agreed upon, and texts tend to differ only in writing style and emphasis of topics. It is usual to tinker with the course, but a major overhaul is difficult to accomplish. Until very recently, the mathematical community was not ready to accept a radical change in the established approach to the subject. If computing is an adjunct to the course, then this addition further crowds an already crowded course. Before the old and proven is replaced with something new, careful consideration must be given to the tradeoffs. Mathematics faculties are not totally resistant to change, but they are, in general, conservative with regard to implementing changes in one of the basic courses of their discipline.

The fact that students were required to program and the necessity to teach the fundamentals of programming lent credence to the question "What do we give up?" and the objection it represents. This objection is still made. In responses to a questionnaire circulated to colleagues at several colleges and universities by Howard Penn of the U.S. Naval Academy, one can find statements such as "Programming is just very time consuming in relation to the academic credit that students get for it." The moral here is that if the instructor does not place a high value on the effort, students will not see the value to the project. Another aspect of the problem is that a student who has to work to get a program to compile will lose sight of the goal of the project and be satisfied with a syntactically correct program that produces output. Little or no analysis is made of the results. Another reply found in Penn's survey was, "A certain number of computer projects is useful, but an overemphasis on computing can become intrusive." This is a delicate balance to establish and maintain.

Other issues that affected the general acceptance of computing in calculus were related to the state of the technology. For the most part, software developed at one institution was not easily transported to other institutions. Software written for one mainframe could not easily be made to run on another. Further complicating the issue, each vendor had its own dialects of the high level languages. Thus a listing of a program had to be carefully reviewed and modified before entering it into a different vendor's computer. An instructor who wished to use computing in calculus generally had to develop materials on site. Conferences and workshops shared ideas, but software was a different issue. Software development required a great deal of time, effort and commitment from the instructor. Consequently, there was no standard model for the introduction of computing. Instructors developed materials that coincided with individual preferences.

A further complication was that most of the computing installations were large mainframes designed to service their entire institutions. Scheduling was a problem, and, in addition to programming skills, users had to learn the ins and outs of the operating system. Facilities for classroom display were expensive and not well developed. This meant that any instruction that required the use of a computer had to be scheduled in a setting other than the standard classroom or without the benefit of hands–on illustrations. Today these problems have been solved by the advent of microcomputers and software developed for them. It is a rare mathematics department that does not have access to microcomputers and units for classroom display of programs and results. These resources are considered in the next section. The hardware and software problems have been alleviated, but there is still a question of the proper use of the tools.

Present Capabilities and Possibilities.

As microcomputers grew in acceptance, the quantity of instructional software for mathematics increased. Some of the individuals who had been producing software on mainframes for calculus produced interactive versions to be implemented on microcomputers. They were joined by others who appreciated the potential of this inexpensive, highly portable tool for the mathematics classroom. The fact that reasonable graphics capabilities were standard on most micros made them even more attractive to software developers and started a new wave of innovative software. Many calculus texts include sections on numerical techniques to be implemented on a computer, and others include a disk with a variety of numerical, graphical, and elementary symbolic manipulation programs related to the material of the text. At the moment theses pieces of software are of uneven quality, and there is no standard collection of programs. However, if this approach to marketing proves to be successful, the marketplace will push toward a standardization of the material published for calculus.

Another low cost option for obtaining software is to use independently developed software. This software is not bound to any book or pedagogical approach. It is designed to be used either for classroom demonstrations or by the student working in a controlled laboratory environment. The user enters necessary data, which may include function statements, intervals, or numeric values. The program then generates graphs and tables related to the data. Instructors may develop several examples prior to class and present a "slide show" to accompany a lecture, or they may interact with the program during the class presentation. The operating systems of microcomputers are so greatly simplified that students need to be given only the simplest of instructions and a brief demonstration to use the software. Three reasonably priced packages of this type are *ARBPLOT* (CONDUIT), *Calculus* (True BASIC), and *The Calculus Toolkit* (Addison–Wesley).

Numerical computing, graphics, and symbolic manipulation constitute the three major areas of present computer use in calculus courses. Many of the good packages combine these three areas. A user may choose a function, draw its graph, find the derived function and plot its graph, print a table of values, and locate critical points for the function, all without having to reenter the function. Thus a complete problem session may be easily presented in a single session with the software package. The instructor will most likely choose to use these capabilities in concert with each other; however, I will discuss these areas separately to emphasize the nature of each.

Numerical Techniques

These provide a bridge between the present software and that developed for the earliest uses of computing in calculus. The major difference is that the student does not do the programming. If the software has been carefully developed, the student does not sit passively observing output. Instead, the student works in an interactive fashion with the program to provide output. While avoiding the need for the student to program, the software must expose the essential nature of the routine and carefully consider the estimate of the error of an approximation. For example, a routine to find a Taylor approximation to a function might require the user to supply the values of the derivatives at the point of expansion, a range of values for the approximation, and bounds on the value of the derivative used in the error estimate. The program could then display the polynomial (preferably using algebraic notation) and the error. A table of values comparing the values of the polynomial and the function across the specified domain, together with the actual errors, can be printed, and, if graphics display is possible, the graph of the function and the polynomial can be displayed.

Structuring a program in this way can increase students' understanding of the numerical technique and the range of its applicability. Students are not forced to deal with the syntax of a programming language, nor are they forced to do a series of tedious homework problems that contain less than one tenth of the information obtained from the software application. The numerical display reinforces the validity of the numerical process, and students learn to develop a healthy skepticism, which is essential for understanding the process. The role of the instructor is very important. The wise choice of a variety of examples that illustrate a number of cases and a careful discussion of the results are the keys to the successful use of this software.

Computer Graphics

The existence of inexpensive computer graphics on microcomputers is a major factor in their use in mathematics classrooms. The adage that a "picture is worth a thousand words" may not be universally true, but it seems to be the case that a good interactive graphics package is worth hours of frustration at a blackboard. There is hardly any subject in a calculus course that the instructor does not wish to illustrate with a graph. Computing can add a dimension of motion to this graphing process. For example, the graph of the derived function can be drawn by plotting the tangent lines to the graph at selected intervals and then drawing the graph of the slopes of these lines. No instructor could or would want to duplicate this process at the blackboard; however, the computer program can do it in a matter of three or four seconds. The graph obtained can provide convincing evidence for a discussion of the relationship of a graph with that of its derived function. The process can be reversed when discussing the antiderivative of a function. Whether this type of illustration is useful depends a great deal upon the skill, judgment, and enthusiasm of the instructor. For example, one can debate whether it is more instructive to see an area approximated by a program using fifty subintervals than it is to see it approximated by the instructor at the blackboard using four subintervals. Perhaps both illustrations should be used. This is a matter for individual taste.

Computer graphics cannot be used to prove a result, but it can be a powerful force for convincing a skeptic. For example, even after seeing a proof of Taylor's Theorem, many have trouble believing that the series

$$x - \frac{x^3}{3!} + \frac{x^5}{5!} - \frac{x^7}{7!} + \cdots$$

represents a periodic function. It is just not intuitively clear! By graphing a high degree Taylor polynomial over the range $[-3\pi, 3\pi]$, the skeptic may be convinced. There is, of course, the temptation to substitute the graphical output for the proof. Even the currently available high resolution graphics is an imperfect medium.

A suggestion has been made [12] that one investigate the concept of slope by examining the actual function magnified in the vicinity of a point in question. What we may be magnifying is not the function but the flaw in the computer's representation of the function. The caveat is that the instructor should not become enamored with the medium and loose sight of the mathematics being illustrated. Visual displays are an aid to understanding, not a substitute for thought.

Three dimensional graphical display was always a problem before we had good computer graphics programs. It is now possible to investigate easily a variety of surfaces in three dimensional space. Most programs require that the user specify the coordinates of the view point and the location of the viewing plane. In the course of explaining the operation of the three dimensional program, the instructor has an opportunity to illustrate a current application of matrix algebra. As microcomputer capabilities and algorithms become more sophisticated, users will be able to rotate surfaces instantaneously and view them from several directions. Our students may become more familiar with these surfaces than their instructors are now. They will not be restricted to viewing them from the only perspective that provides for an easy drawing.

Graphics packages for use in calculus vary, and their claims do not always match their performance. It is necessary to test them with a representative number of functions prior to purchase. The test should determine ccc the package handles special cases, how it allows the user to change parameters, and its general ease of use. Another factor to consider is the resolution of the display which is, of course, limited by the hardware being used. However, some software authors work within the hardware confines better than others. Each package must be reviewed from the perspective of its proposed use. Obviously the software should be designed to fit the use, not the opposite. The article by Cunningham in this volume gives guidelines for software evaluation.

Symbolic Manipulation Packages

These packages are receiving a great deal of attention from calculus instructors. They are advocated by some as providing a springboard for the development of new and radical approaches to the calculus. Among other things, these packages do symbolic differentiation, integration, and solutions of equations using exact arithmetic. The inputs are strings of symbols, not just decimal numbers. The programs operate on the strings using the rules of, say, the calculus. For example, if one were to input INT$((x-1)/(x+1),x)$; the program would most likely output $x - 2\ln(x+1)$.

The primary packages for symbolic manipulation are MACSYMA, Maple, muMath, Reduce, and SMP. Most of these require a mainframe or a high end minicomputer. muMath was designed to run on a microcomputer, but it has limited capabilities and a rather unsatisfactory output format. A version of Maple is available for the Macintosh Plus, a rather powerful microcomputer. Some packages designed for other purposes, such as True BASIC's *Calculus*, have very limited symbolic differentiation routines which are unsatisfactory for general use. John Hosack's paper in this volume discusses the instructional uses of symbolic manipulation packages. The discussion here is limited to their use in a calculus course.

John Kemeny once stated in an article in the *Christian Science Monitor* that computing frees us from the drudgery of arithmetic. However, messy arithmetic is not the only drudgery found in a calculus course. Many times the need to find the derivative or integral of a function introduces drudgery into a problem designed to illustrate some concept other than the mechanics of these processes. Symbolic manipulation packages can do such processes with relative ease. Thus, they free the student to concentrate on the essential steps of problem solving.

There is a valid concern about when to introduce symbolic manipulation programs to students in a calculus course. It is argued that students need to wrestle with the operations of solving equations, symbolic differentiation, and integration to appreciate their underlying concepts. It is also argued that working with tedious mechanical operations can force the students to develop skills, such as pattern recognition and the observation of more general rules, that will be valuable in courses beyond the calculus. For example, the recognition of which technique of integration to apply to a function and how subtle changes in the function affect the choice of a technique develops pattern recognition and search techniques that may have valuable transfers in other mathematics, chemistry, computer science, or physics courses. If one does not struggle with the Taylor expansion of $\arctan x$ about $x = 0$, how does one reinforce the relationship of this expansion and that of $1/(1-x)$? A complete discussion of this topic might fill a volume of its own, and I will not address the concern. My purpose is to discuss ways in which symbolic manipulators can be applied within the calculus course.

One obvious use of symbolic manipulation packages is for curve sketching exercises. Much time is spent in finding the derivatives of the function and in the determination of critical points. By the time the student gets to the point where it is possible to use the information obtained to sketch the curve, a lot of hard work has already been done. The student rushes through the main part of the exercise with the feeling that the "important" parts have been done. This is a poor use of students' time. The use of a symbolic manipulation package forces students to concentrate on the topic of the assignment. The same observation applies to the use of symbolic manipulation for assignments dealing with applications of the calculus.

Symbolic manipulation packages are especially useful for instructors who wish their students to apply the *methods* of calculus. That word was carefully chosen; methods are not techniques. Unfortunately, by design or default, many calculus courses emphasize only techniques. A classic lament of calculus students is that they can do the calculus, it is just those word problems that give them difficulty. The removal of the manipulation phase of solving the problem forces concentration on the processes of setting up the problem and analyzing the results of the manipulation to solve the problem. Hosack and Small [3] state that the availability of a computer algebra system (symbolic manipulation package) allows one to develop a basic approach to problem solving founded on the following questions:

1. What do I need to know?
2. What can I tell by inspection?
3. What are the possibilities suggested by what I know?
4. What should I do next?

The approach is not new, but the possibilities that exist as a result of symbolic manipulation allow the process to flow smoothly without being interrupted by the need to apply a technique that is not in the main stream of the flow.

Another advantage of the symbolic manipulation programs is that they enable the instructor to ask open–ended questions with some confidence that students will attempt to find answers. For example, one might ask about the antiderivatives of expressions of the form $(a + bx + cx^2)^q$: What happens when one varies a, b, c, and q? Questions of this sort are simply beyond the capabilities of most first year students who do not have access to symbolic manipulation packages. They would be overwhelmed by the number of integrals they would need to compute. The availability of these packages extends the "what if" capabilities of our students.

Thoughts About the Present Situation

Certainly the advent of microcomputers does not mean that there will be a rush by all calculus instructors to use computing in calculus. There are factors other than the availability of hardware and software affecting this issue. Many of us are comfortable with the calculus course we have been teaching and are not eager to devote a great deal of time and effort into changing this course. The syllabus for the course is well established and known by virtually all of our colleagues across the campus. Many departments rely on the calculus course to cover certain topics and insist that those topics remain in the course. Mathematics departments have little freedom to change the course unilaterally. It is a service course used by many departments for a variety of purposes.

Computing takes additional time. It is difficult to find time in an already crowded syllabus. If computing is simply added to the existing course, the project is doomed to failure. The instructor must make the use of computing a prominent part of the course presentation. This requires a major effort. Textbook publishers tend to publish texts that conform to a proven formula. New approaches have a limited appeal and represent a risk. Complicating this is the fact that the text may require software that is not distributed by the publisher. These facts leave the dedicated and innovative instructor without easy access to resources for implementing change.

Although these two major impediments to computing in calculus exist, it is not time to write the obituary for the effort. We have seen that there have been major improvements in the software tools, and the price of equipment to implement the tools is reasonable. On the other hand, the existing calculus course is under attack [2]. It is becoming apparent that we are not reaching all of the students for whom the course is intended. There are convincing arguments that we are not reaching a large percentage of the students we have in our calculus courses. Many of these students find the material dull and see little, if any, justification for taking calculus. This is a sorry situation for a course that occupies a central position in the undergraduate curriculum. The use of computing is not offered as a cure for a troubling and deep seated problem. It can, however, provide a means for accomplishing a restructured approach to the calculus that preserves the espoused goals of the course.

Some Course Options.

This section presents a general discussion of alternative approaches to teaching the first year course and a brief mention of some possibilities for multivariate calaculus. It is neither exhaustive in its listing of options nor detailed in its discussion of any one option. My intent is to entice the reader to consider the possibilities for change enabled by the existing computing resources.

A Laboratory Calculus Course

This course has the same design as the laboratory course in the sciences. Students meet with their instructor for lecture and discussion, and there is a scheduled laboratory period. During this period students run experiments based on the material presented in the other portion of the course. Some sample laboratory projects are: an examination of convergence properties; an exploration of the definitions of the derivative and definite integral; the relationships among the graphs of a function, its derivative, and antiderivative; development of polynomial approximations to a function and comparison of the graphs of the function and its approximants. These projects are most effective if the student does a minimal amount of programming and results are accompanied by graphical displays. A crucial part of each lab is a brief written report that analyzes the results of the experiment. The purpose of the lab is to have the student think about calculus, not just "do" calculus.

The lecture–discussion portion of the course also includes demonstrations. The classroom setup must allow for easy access to a computer with a display that is easily viewed by all of the students. Demonstrations can be used to motivate topics to be introduced or to aid in solving problems. The important point is that the spirit of the course is one of using a powerful computational tool to help with the understanding of calculus. A side effect of the demonstrations is that the instructor becomes a role model for the use of the computer in conducting an investigation of a topic in calculus.

Much time must be spent in the preparation of material for this course. In addition to choosing a text, the instructor must carefully choose the software and test it. It is then necessary to develop laboratory projects and write a lab manual. These are all time consuming chores, but they are absolutely necessary if the course is to be a success.

The natural objection is that this course makes calculus something other than a mathematics course in the traditional sense, that it may not be appropriate for mathematics majors. It is true that this is not a traditional mathematics course, but then the calculus course as it is presently being taught is different from any of the advanced courses in the mathematics curriculum. We do not present calculus in the standard Definition, Theorem, Proof format. The proofs of the deeper and more interesting theorems are avoided, and we stress the applications of the theorems more than the theorems themselves. In a sense, the first year calculus course is a mathematical engineering course. The computer laboratory is a forum for students to test their intuition and make conjectures.

A Survey of Calculus Course

This is the course designed for non–mathematics majors. The students who usually take this course are social science and humanities majors. The goal of the course is to provide students with a general understanding of the calculus and methods of applying these concepts. It is also the course where we miss our goals by the widest margin. Students leave the course with the ability to find derivatives and antiderivatives for a limited class of functions. Since students are restricted in the class of functions they consider, the applications tend to be shallow and transparent when viewed from the disciplines from which they are drawn. The course proposed here removes these restrictions.

This is an interactive course. Students and instructor experiment, discuss their observations, and develop theory. Although students will do some of the traditional exercises to learn techniques, there is a heavy reliance on graphic displays and symbolic manipulation packages. Students learn a minimal number of techniques to gain a general understanding of how the symbolic manipulation package implements its routines. We certainly do not want to surround the symbolic manipulator with a black box mystique.

This course stresses applications and analysis of the results of applying calculus. A central part of the course consists of projects that require the application of calculus material to problems in the student's area of interest. In this respect the course is similar to an elements of modelling course. Each project is generated by a student under the direction of the instructor and a faculty member in a department related to the student's interest. An alternative approach to the projects is to have faculty from other departments submit interesting problems that would be amenable to solution by the methods of the calculus. The instructor would then edit these problems and allow students to choose from among them.

Since there are virtually no materials developed for this course, the instructor needs to develop careful lecture materials and lecture notes. The software chosen for this course must be easy to use and the illustrative examples carefully chosen. It is important that the examples be neither trivial nor involved. One of the goals is that students develop a feeling for numeric and non–numeric data. When the student is confronted with data or a statement about a phenomenon, it should be natural to ask: What relationship exists among the variables present in the data? Is this relationship a function? What do I know about this function? These are questions for which the calculus can supply answers. Furthermore, the software tools used in this course must provide help in obtaining these answers.

This course is also a radical departure from the existing non–majors calculus course. The reason for redesigning the course is to show students in essentially non–mathematical disciplines that calculus has the potential for analysis of phenomena in several fields. It is not the exclusive property of natural scientists, engineers, and mathematicians.

An Algorithmic Calculus Course

This approach was used by the CRICISAM project in the late 1960's. The basis for the presentation of the calculus is the notion of convergence of sequences. One first discusses the meaning of convergence of a sequence $\{z_n\}$ to zero. The limit concept, continuity, and differentiability are taught using this notion. For example, F is continuous at $x = a$ if $\{F(a+z_n)\} \longrightarrow F(a)$ for all sequences $\{z_n\}$ such that $\{z_n\} \longrightarrow 0$. This approach leads naturally to algorithmic presentations of proofs of theorems and for setting up applications. Although the course is not a discrete mathematics course, data for examples are given as discrete terms of a sequence. Proofs of theorems are given in terms of constructions.

One advantage of this approach is that it easily lends itself to a presentation of functions defined on $\mathbf{R}^n$. There is no reason for the instructor to be confined to a discussion on the real line or a graph displayed on the Cartesian coordinate plane. The idea of convergence of a sequence is very extendible. The instructor using this approach will find packages containing programs for displaying surfaces and a matrix calculator invaluable aids.

This approach is not the standard fare for most mathematics departments. It may be of interest to departments working on a sequence that combines discrete mathematics and calculus as separate courses during the first year. An instructor considering this course should review the CRICISAM notes [11] and Bishop's book on a constructive approach to analysis [1]. John Hosack and Donald Small have written a set of course notes for the course they teach at Colby College. These notes are at present in unpublished form [8].

A Combined Calculus and Discrete Mathematics Course

This course is designed in response to the current discussion about the importance of calculus within the first year of the mathematics curriculum. It is not a common course, and it is still under development. One project is under way at George Mason University under the direction of Stephen Seidman and Michael Rice. The goal of this course is to provide students of computer science with a feeling for the connection between discrete mathematics and calculus. It is an energetic project that has a perspective of the broad spectrum of mathematics, which is not limited by looking at only discrete techniques or only continuous techniques.

From this perspective, results from discrete mathematics can give insight into results in continuous mathematics and vice versa. An example of the former is that the behavior of the difference operator can be used to motivate the properties of the differential operator. An illustration of the latter case is that results concerning the term by term differentiation of a series within its interval of convergence can be used to obtain the value of an unfamiliar series, such as the series whose terms are $k2^{-k}$. A very effective tool in this course is a symbolic manipulation program that can, in addition to doing symbolic differentiation and integration, calculate partial sums of series and evaluate limits involving symbolic constants. This tool can keep students from becoming bogged down in long, messy calculations.

The method of presentation of this combination of topics is, quite obviously, algorithmic. The complete presentation is a sequence of two courses; the first concentrates on functions and function behavior, the second on structure. Seidman and Rice have included course outlines in [7].

Beyond the First Year: Multi–Variable Calculus

If computing plays a central role in the first calculus course, it should continue to be featured in subsequent courses. Obvious choices for projects are those that parallel projects in the first course, for example, evaluation of limits along different paths, evaluation of directional derivatives and gradients, and sums for double and iterated integrals. These routines are found in most numerical packages.

Other projects require matrix arithmetic. This can be done with a symbolic manipulation package that has routines to add, subtract, multiply, and invert matrices, or with one of the matrix calculator packages. Using one of these packages eliminates the need to do lengthy calculations and thus frees the user to observe that many of the results of multivariable calculus are obvious extensions of those of single variable calculus. In the best case, the student may even make conjectures about extensions not yet covered in the course.

Since symbolic manipulation packages and matrix calculator packages can solve systems of equations, they can be used to find optimum values for constrained functions of several variables. Hosack and Small use the following illustration [9].

Find the extrema of the function

$$f(x,y,z,t) = x^2 + 2y^2 + z^2 + t^2$$

subject to:

$$g_1(x,y,z,t) = x + 3y - z - t - 2 = 0,$$

$$g_2(x,y,z,t) = 2x - y + z + 2t - 4 = 0.$$

After defining the Lagrange function, the student enters the appropriate set of partial derivatives set equal to zero and the computer returns

$$x = 67/69,\ y = 2/23,\ z = 14/69,\ t = 67/69,$$

and the constants introduced in the Lagrange function, –26/69 and 18/23. The student has done the important parts of the problem without spending time solving a system of equations.

Graphics illustrations in this course are extremely useful for building intuition about functions of more than one variable. Ease of use is an important factor in choosing a program for this purpose. The True BASIC *3-D Graphics* package has a sample surface plotting routine that is easily modified to illustrate different surfaces.

These are a few suggestions for computer use in the second calculus course. The important consideration is that the momentum built in the first year calculus should not be lost when the student advances. This adds to the idea that the computer is a powerful aid in the problem solving process.

Summary.

Calculus with computing is entering a new era brought about by the availability of inexpensive hardware and subsequent activity in the development of software. Whether it will fare any better in this new era than it did in its old environment remains to be seen. Many problems, such as scheduling, coordinating the presentation, and preparing interesting and challenging examples, still exist. On the other hand, the logistics have been made easier and the need to program eliminated.

One factor present now but not earlier is that there is a generally felt need to reconsider the way in which calculus is taught. Many of the revised presentations of the calculus feature the use of computing for graphics illustrations or symbolic manipulation. It is not coincidental that the availability of these tools and the cries for change in the calculus course are occurring at the same time. The tools very well may have spurred the mathematical community to think about its presentation of this course. Computers and computer science have given us a clientele with a different perspective from that of previous years. It is apparent that there will be further developments in the use of computing in calculus.

Acknowledgement.

The preparation of this article was greatly aided by the following individuals who contributed information and materials in response to a request in *Focus*:

James Burling, State University of New York at Oswego;
Howard Penn, The United States Naval Academy at Annapolis;
Stephen B. Seidmen and Michael D. Rice, George Mason University;
Donald Small, John Hosack, and Kenneth Lane, Colby College.

REFERENCES

1. Bishop, Errett. **Foundations of Constructive Analysis**. New York: McGraw–Hill, 1967.

2. Douglas, Ronald G. (ed.). **Toward a Lean and Lively Calculus**, MAA Notes No. 6. Washington, DC: Mathematical Association of America, 1986.

3. Hosack, John M., Kenneth Lane, and Donald B. Small. "Report on the Use of a Symbolic Mathematics System In Undergraduate Instruction." **SIGSAM Bulletin 19** (1985), 19–22.

4. Leinbach, L. Carl. "Calculus as a Laboratory Course." **Proceedings of the Second Conference on Computers in the Undergraduate Curricula**. Hanover, NH: Dartmouth College, 1971, pp. 13–17.

5. Leinbach, L. Carl. **Calculus with the Computer: A Laboratory Manual**. Englewood Cliffs, NJ: Prentice–Hall, 1974.

6. Porter, Gerald J. "A Computer Assisted Approach to Integrals via Jordan Content." **American Mathematical Monthly 77** (1979), 417–419.

7. Porter, Gerald J. "The Computer as a Teaching Aid in Mathematics." **Proceedings of the First Conference on Computers in the Undergraduate Curricula**. Iowa City, IA: University of Iowa, 1970, pp. 4.17–4.20.

8. Small, Donald B., John M. Hosack, and Kenneth Lane. **Calculus of One and Several Variables: An Integrated Approach**. Waterville, ME: Colby College, 1983.

9. Small, Donald B., John M. Hosack, and Kenneth Lane. "Computer Algebra Systems in Undergraduate Instruction." **The College Mathematics Journal 17** (1986), 423–433

10. Smith, David A. **Interface: Calculus and the Computer, 2nd ed.** Philadelphia: Saunders College Publishing, 1984.

11. Stenberg, Warren, Robert J. Walker, et al. **Calculus, A Computer Oriented Approach** (Parts 1–5). Tallahassee, FL: The Center for Research in College Instruction in Science and Mathematics, 1968.

12. Tall, David. "Visualizing Calculus Concepts Using a Computer." **Proceedings of a Conference on The Influence of Computers and Informatics on Mathematics and Its Teaching**. Strasbourg, 1985, pp. 291–295.

A Minimalist Approach to Computer Use in Mathematics

David A. Smith

The minimalist approach to education can be stated simply: *Don't do too much for the student.* (The sentence may be read with emphasis on any of the following: "do," "too much," "for," or "student." Try it!) An expansion of that idea into the realm of computer support for education might look like this: Do enough to avoid unnecessary frustration, but don't do so much that the learning process is shortcircuited. (On reflection, that maxim may have broader application than just "computer support," but let's stick to the topic at hand.)

The actual implementation of the minimalist position is highly time–dependent. There was a time when it seemed necessary to become a programmer in order to make *any* use of a computer. In that context, the frustration–avoidance devices were things like good documentation (rarely available), supplementary handouts (almost always required), student consultants, and faculty presence in the "lab." It was often necessary to devise and present a working subset of a language and an approach to learning that language that was quite different from that in the programming textbooks. "Not doing too much" may have meant only "don't provide finished programs that solve the problems — let the students write their own."

At present there are obviously far more computer users than there are programmers; the "early days" of required programming are behind us. In their place we have a context in which avoidance of frustration is expressed in terms of a "friendly user interface." Our tools for computing now include a relatively small number of relatively standardized hardware configurations and operating systems and a vast array of software environments that "support" a wide range of user categories.

My purpose in this article is to describe an application of the minimalist principle at Duke University that has served us well over a decade that starts in the "early days" and extends to the present. At the end of the article I will comment on other ways the same concept might be implemented and on other courses in which it would be useful to do so, including some that are represented by other contributions in this volume.

MATHPROGRAM, a spreadsheet for mathematics.

F. J. Murray was the first of my colleagues to expound the minimalist principle, and his idea for implementing it came to be called MATHPROGRAM. I will describe briefly the concept behind the original MATHPROGRAM and its successors; further details of the program and our first three years of experience with it may be found in [2].

MATHPROGRAM is a simple but general sequence tabulator, capable of tabulating and/or graphing any specified number of terms of up to seven interrelated sequences, each defined either recursively or explicitly. In effect it is a highly specialized spreadsheet (N rows by seven columns), streamlined for use in calculus and other mathematics courses. Unlike conventional spreadsheets, the formulas are entered globally for columns rather than in individual cells. To use this program, the student enters definitions of any functions involved, formulas for advancing the sequences used from step $n - 1$ to step n, starting values at step 0, and possibly some constants. Thus the extent of programming knowledge required is the ability to write an assignment statement, the one programming concept most like writing a mathematical formula in conventional notation. The program is general enough to handle function tabulation, root finding, numerical integration, summation of series, and many other concepts frequently encountered in a calculus course. (Examples to show how it does all these things will be given in the next section.)

While our principle applications have been to the study of calculus, we have also made effective use of various versions of the program in differential equations, complex analysis, advanced calculus, number theory, and numerical analysis. I will say more about discrete mathematics below; if we had such a course in our curriculum, some version of MATHPROGRAM would be a primary tool there also.

The sense in which MATHPROGRAM satisfies the minimalist principle is this: It does not short–circuit learning about numerical methods and their interplay with calculus, because students are involved with each numerical method as it comes along. They have to find out *what* formulas to enter into the program and use those formulas correctly. They also have to *interpret* the tabulated or graphical output and answer both specific and open–ended questions about that output. On the other hand, much of the frustration of the "write your own programs" environment is avoided. If the program does not run or produces the wrong answers (it is the student's responsibility to detect the latter), the bugs can occur only in the handful of lines they have entered, and the types of bugs are limited to the (mostly mathematical) details of assignment statements. Students are shielded from essentially all the non–mathematical details of computer use.

In another paper in this volume John Hosack endorses the principle of removing non–mathematical frustrations, but he also states, "... few instructors believe that the pedagogical advantages of implementing numerical algorithms are worth the time, especially when the emphasis of the course remains on symbolic algorithms." I have long believed there are worthwhile pedagogical advantages in the use of numerical algorithms in parallel with the symbolic ones, but I endorse the implication in Hosack's statement (explicated elsewhere in his paper) that the time has come to change the emphasis of the calculus course. I have no quarrel with the use of computer algebra systems as tools from the outset of college–level mathematics. In fact, I think students should learn to do both symbolic and numerical calculation with the best tools available (i.e., with minimal frustration), provided they also understand the mathematics underlying their powerful tools. It happens that numerical computation can be done with what we now see as "primitive" tools, while symbolic computation still requires relatively sophisticated tools. Hosack and I are both advocating getting the most out of the tools available, whichever end of the technological spectrum you happen to be closer to, without short–circuiting the learning process.

There have been many versions of MATHPROGRAM over the years; only the time–sharing BASIC and Applesoft versions were ever published [3]. Early versions had no graphics capabilities, but the advent of microcomputers forced us to reevaluate what constituted "too much" help for students in that area. In particular, the prospect of inexpensive, high–resolution graphics eventually in the hands of every student (that "eventually" is now!) suggested that graph paper would become as obsolete as the slide rule. Thus the Applesoft version and equivalent versions for the Commodore 64 and IBM PC have the ability to graph any sequence as a function of any other, along with a menu–driven user interface to remove more of the frustrations for computer novices. (The IBM version is our standard tool for calculus now.)

How MATHPROGRAM works.

Here are some examples (all from the calculus course and in standard mathematical notation) to demonstrate the flexibility of the MATHPROGRAM concept.

Function tabulation. If a function $y = f(x)$ is given, and one makes a table of values (in preparation for graphing, say), each of the columns (headed by x and y) is a *list* of numbers, i.e., a (finite) sequence. Any computer graph of the function starts with such lists, usually with equal spacing between the x's; why shouldn't students see and use the lists themselves? For graphing with m subdivisions of the interval $[a,b]$, the key formulas are:

$$\Delta x = (b - a)/m,$$

$$x_n = x_{n-1} + \Delta x,$$

$$y_n = f(x_n).$$

The first of these is a "Start" formula (executed only once), and the others are "Advance" formulas, one recursive, the other explicit. (The recursive advancing of x_n could also be written explicitly, and this would avoid a possible accumulation of roundoff error:

$$x_n = a + n\,\Delta x.$$

No harm is done if students are allowed to discover this themselves.) The user also provides Start formulas

$$x_0 = a,\ y_0 = f(a),$$

a definition for f, and values for a and b.

Actually, this is the default configuration in all versions of MATHPROGRAM because it is so valuable and is used so heavily at the beginning of the course. With the graphic capability of the micro versions, there is an instant demonstration of inverse functions (or relations): Graph y vs. x, then x vs. y. Once the basic tabulation idea is understood, it can be replicated with other letters to do simultaneous tabulations and/or graphs of two or more functions, say f, f', and f'', or f, g, $f \circ g$, and $g \circ f$.

Finding roots. Function tabulation provides a crude but effective way to locate a zero of a function to any desired accuracy: Tabulate with unit spacing; where f changes sign, you have located a root between consecutive integers, say c and $c + 1$. Tabulate again on $[c, c+1]$ with $m = 10$. You now know the location between consecutive tenths. Change the end points again, and repeat. Each "interval tenthing" tabulation gives one more decimal place of the answer. This may appear to be tedious, but only the constants a and b change between runs, and finding one root goes pretty quickly. On the other hand, if you want to find a lot of roots, there has to be a better way.

The usual interval–halving technique can be implemented if you are willing to add IF–THEN statements to the programming repertoire. A better way to proceed is to go right to Newton's method, an obviously important application of the linear approximation concept that is central to differential calculus. By this time there is a MATHPROGRAM configuration that tabulates f and f' (possibly more), with an advancing formula of the form

$$z_n = f'(x_n).$$

Now change just *one line* in the tabulation routine:

$$x_n = x_{n-1} - y_{n-1}/z_{n-1}.$$

That's Newton's method! The initial guess (easy to get because you have already tabulated the function) is inserted as the value of a, and the other constants, b and Δx, are simply ignored. Having just done tabulations for graphing with 50 or 100 points, students will start doing Newton's method with large values of m. It doesn't take long for them to discover that nothing changes after the first few steps — that's a "convergence experience."

The Fundamental Theorem of Calculus. The first step in teaching the integral is usually to introduce summation notation and technique. In MATH-PROGRAM that means the generic summation formula

$$\text{sum}_n = \text{sum}_{n-1} + \text{term}_n,$$

with initialization $\text{sum}_0 = 0$. Riemann sums are introduced and calculated by taking term_n from a function tabulation and multiplying by Δx (which can be done either before or after summation). In particular, the most effective and important Riemann sums (those of the Midpoint Rule) are given by

$$\text{term}_n = f(x_n - \Delta x/2),$$

$$I_n = \text{sum}_n\, \Delta x.$$

Similarly, when we are ready for Simpson's Rule, we can write

$$z_n = f(x_n - \Delta x/2),$$

$$\text{term}_n = y_{n-1} + 4z_n + y_n,$$

$$I_n = \text{sum}_n\, \Delta x/6.$$

Only the last entry, I_m, is of interest for the definite integral problem, but the I column is obviously a tabulated function. What function? Well, it's a good approximation to that seldom–understood antiderivative whose existence is assured by the first part of the Fundamental Theorem, namely $F(x) = \int_a^x f(t)dt$. We can demonstrate what the Fundamental Theorem really says by adding one more formula to differentiate F numerically:

$$\text{diff}_n = (I_n - I_{n-1})/\Delta x.$$

That gives a very good approximation to $F'(x_n - \Delta x/2)$, which should be the same thing as the tabulation in the z_n column. The results will be *close*, not exact, but the graphs of diff_n and z_n as functions of x_n will be identical.

The most important aspect of this example is that we can demonstrate in a very tangible way the truth of the Fundamental Theorem for *any continuous function*, not just for those whose antiderivatives we can find in closed form. Indeed, we can do it *before* any discussion of techniques of integration. The same formulas can be rearranged a little to demonstrate the rest of the Fundamental Theorem by differentiating first, then integrating. The details are left as an exercise — be careful with the constant of integration.

Sums of series. Most calculus books pay a lot of attention to the question of whether a series converges (a relatively minor problem in practice), very little attention to what its sum is if it does converge (the whole point of representing things by series). Suppose, in the course of discussing convergence of alternating series, you told your students that $\sum_{n=1}^{\infty} \frac{(-1)^{n+1}}{n^2} = \frac{\pi^2}{12}$. Why should they believe anything as absurd as that? And how would you convince them? Our generic summation formula can be put to work again with

$$\text{sign}_n = -\text{sign}_{n-1},$$

$$\text{term}_n = \text{sign}_n / n^2,$$

and the proper start formulas. This series converges much too slowly to be evaluated on a calculator, but not so slowly that roundoff errors become a problem on the computer. If you ask for the 200th partial sum, your error will be no worse than $1/201^2 \approx 2.5 \times 10^{-5}$. However, you can do much better and in the process convey a better understanding of alternating series. It is easy to see (either formally or by looking at a picture of partial sums) that the sum of a convergent alternating series is closely approximated by adding *half* the next term to your current partial sum. It is almost as easy to show that the error in this "corrected" partial sum is bounded by half the absolute value of the sum of the next two terms (which are nearly equal and of opposite signs). Thus we may supplement the formulas already given by

$$\text{corsum}_n = \text{sum}_{n-1} + \text{term}_n/2,$$

$$\text{error}_n = \left|\text{term}_n + \text{term}_n\right|/2.$$

[If the implementation of the sequence tabulator does not permit looking ahead to an $(n+1)$th term, the subscripts on the right can be lowered by one, and the error bound for the *current* corsum is then in the *next* row of the printed table.] The 50th corrected sum is about 0.822471, with error not more than 8×10^{-6}. The actual error is less than half that, so comparison with a calculator value for $\pi^2/12$ looks pretty good. Furthermore, to sum any other alternating series, the only formula that needs to be changed is that for term_n. (This and several other techniques for efficient and instructive summation of series are discussed in detail in [4], pages 212–228.)

Other ways to implement the minimalist concept.

MATHPROGRAM predates VisiCalc; we used it for years before realizing it was a spreadsheet. Clearly what we are doing with MATHPROGRAM could be done with any "conventional" spreadsheet, although effective use in the present graphics–oriented environment requires that there be an easy and flexible way to graph the results of the calculations. For some very popular spreadsheet programs (AppleWorks, for example), that requires an extra–cost add–on program. In my opinion, having to "replicate" a formula to make it global for a column is an unnatural operation, one that is unrelated to recursive or explicit definition of sequences. However, it is an easy operation to learn and do with any reasonable spreadsheet. The notation of spreadsheet cell names and formulas is not as close to mathematical notation as programming notation is, but it too is easy to learn.

The idea of using a business–oriented spreadsheet to do mathematical calculations is not new. For example, Arganbright [1] has explored the use of such a spreadsheet for many of the same calculations we do with MATHPROGRAM, plus some we have only thought about. (We have implemented versions with functions of more than one variable and with complex arithmetic, but not with matrix operations.) Is this a reasonable thing to do with large numbers of students? Clearly the answer is "no" if it means an expense of several hundred dollars per student or per computer. However, if your campus or department already has a site license for a good spreadsheet (or has enough reasons for getting one), why limit its use to secretaries and administrators? Software publishers now offer "educational" versions or special bulk purchase or site license arrangements on popular spreadsheet programs to encourage their use for teaching purposes.

Furthermore, there are good spreadsheets among the "freeware" and "shareware" (often public domain) offerings available for major lines of microcomputers. For example, for the IBM PC and compatibles there is ExpressCalc, which is copyrighted, but permission is given for free copying for educational purposes, and the initial cost is quite modest. To get graphical output, you also need ExpressGraph, which is available on the same basis, and having to switch programs to get graphs is an added complication, but not a disabling one. (For details on both programs, contact Expressware, P. O. Box 230, Redmond, WA 98073.)

I have a more radical proposal for future implementation of the minimalist principle: Do it in Logo! I have only begun to explore this idea, but it seems natural to use a procedure–oriented, recursive, list processing language with easy–to–use graphics built in. The operations of MATHPROGRAM (and substantial generalizations thereof) are all easy to code in Logo, and students using such a "microworld" would have the tools for expanding the environment far beyond the basic capabilities provided for them. In effect, they would design and build their own microworlds for calculus, discrete mathematics, differential equations, statistics, and perhaps many other courses or combinations thereof.

There is a cost factor here, too; it may not seem as obvious that everyone should own and use Logo as it is for, say, some version of BASIC or Pascal. However, the articles by King and by Jones in this volume show that the range of college–level applications of Logo is quite broad, so a departmental site license may turn out to be a real bargain. Capabilities of different Logo implementations vary widely, but the language is so easy to extend that missing features (e.g., most of the elementary transcendental functions) are usually available as "utilities" provided with the system or in the public domain.

The structural virtues of Logo are shared by APL to a considerable extent; indeed, one of the more powerful implementations of MATHPROGRAM, no longer in use, was written in APL. APL has greater numerical computation capabilities and more built–in operations for function evaluation and matrix manipulation, but it lacks the native graphics capabilities of Logo, and it is much harder for a computer novice to use effectively. This last is not a problem if the routines are to be provided to students, but it will not be as easy for them to expand their microworld as it is with Logo. For a specific APL implementation that is fully consistent with the minimalist principle, see Helzer's contribution to the Linear Algebra Panel Report in this volume.

Applications of the minimalist principle.

Whether implemented by a general (business oriented) spreadsheet or a specialized one, and whether the software is written in Logo, APL, or some less suited language (such as BASIC), the minimalist concept can be applied in a wide variety of courses. Two obvious advantages of this: A department can cover a lot of ground with a single piece of software; and students will know how to compute in all those courses when they have learned the operation of just one program.

For example, essentially all of the discrete mathematics topics discussed in the papers by Siegel and by Sandefur and Vogt in this volume lead naturally to recursive sequence computations. (Matrix and vector operations would have to be provided for some topics; while they are not in MATHPROGRAM, they are certainly included in the concept.)

My vision for the future of lower division college mathematics and for the resolution of the current debate about "continuous vs. discrete" mathematics is that we will realize that these are not two subjects but one. We already have a lot of discrete mathematics in the calculus course (numerical approximation methods, Riemann sums, difference quotients, Taylor polynomials, sequences and series), but we always see these things as steps on the way to something else, as approximations to the "real" mathematics. It would be healthier to acknowledge that there is a constant interplay between the continuous and the discrete, that each can approximate the other, that either can be the problem we really want to solve.

In this spirit, I have a confession to make. My demonstration of the Fundamental Theorem of Calculus, outlined above, is *outrageously* good. The errors after integrating by the Midpoint Rule and differentiating by symmetric differences are much smaller than one would have a right to expect on the basis of the error estimates for those methods. The reason for this is that the Midpoint Rule and symmetric differencing are inverse operations in exactly the same sense that integration and differentiation are. What is really being demonstrated is the Fundamental Theorem of Discrete Calculus, which says nothing more than "addition and subtraction are inverse processes." The proof is just as obvious as that, but in the limit it becomes the "real" Fundamental Theorem. By sublimating the discrete aspects of the subject, we are missing opportunities for drawing on the arithmetic experience of our students to gain substantial insights into the nature of the "limiting case." Of course, easy access to substantial computations is the key to realizing those opportunities.

This interplay between the continuous and the discrete extends into the realm of differential (and difference) equations. For example, Sandefur and Vogt observe that the closed form technique for solving linear difference equations with constant coefficients exactly parallels the corresponding technique for differential equations. All of the simulations and other computations described by Danby in his contribution to the Panel on Differential Equations can be carried out with MATHPROGRAM or with any spreadsheet with graphics capabilities.

Other papers in this volume suggest additional applications of the minimalist principle. I have already mentioned Helzer's linear algebra course as an example. Arganbright [1] gives examples of the use of a conventional spreadsheet in linear algebra; the matrix calculators designed by and described herein by Herman, Orzech, and Anton may all be viewed as "specialized spreadsheets" for linear algebra. Their authors might prefer the hand–held calculator paradigm, with "registers" capable of holding vectors and matrices and "operation keys" capable of things such as matrix multiplication and inversion. But spreadsheets are, after all, an automated expansion of the capabilities of calculators to handle problems for which it would be too tedious to repeat all the operation keystrokes.

The papers on statistics by Gordon and on probability by Snell and Finn suggest another area of application. Statistical calculations are routinely done on spreadsheets, and statistical analysis software may be viewed as another form of specialized spreadsheet. To carry out simulations in this context one needs to provide routines that simulate the important probability distributions; these may be built up from the random number generator provided with the system. Snell and Finn make a case for clever uses of graphics in specialized simulation programs, some of which would go well beyond the capabilities of the typical spreadsheet, and I would not argue with that.

I could describe applications to courses in number theory, abstract algebra, complex analysis, advanced calculus, numerical analysis, and so on, but the real points are these: (1) Some of these subjects are inherently discrete, and in the others there is an interplay between discrete and continuous, not always recognized. (2) Many of the computations in these areas for which a computer is appropriate or required are repetitions of steps that could be done on a simple calculator; thus the spreadsheet metaphor is all that is needed to accomplish the computational tasks.

The single loop (or single tail–end recursion) concept implemented in MATHPROGRAM is, in one sense, at the opposite end of the software sophistication spectrum from the computer algebra systems discussed by Hosack. And yet, the dominant metaphor in proposed educational uses of CAS's is that of the calculator with keys for operations such as differentiation, integration, and matrix inversion. Such a calculator now exists in a hand–held package with Hewlett–Packard's release this year of the HP 28C. These ideas have a way of coming full circle.

At Duke our present interpretation of the minimalist principle is this: Don't do too much *mathematics* for the student who is learning mathematics, but do everything possible to allow the student to focus entirely on that goal. In particular, one should make it as easy for the computer novice to use the computer as an effective tool as it has become, say, for secretaries to use word processors.

REFERENCES

1. Arganbright, D. E. "The Electronic Spreadsheet and Mathematical Algorithms." **The College Mathematics Journal 15** (1984), 148–157.

2. Barr, R. C., T. M. Gallie, Jr., M. J. Hodel, R. E. Hodel, F. J. Murray, and D. A. Smith. "A General Problem Solver for Mathematics Courses." **Proceedings of the Ninth Conference on Computers in the Undergraduate Curricula**, University of Denver, 1978, pp. 23–35.

3. Barr, R. C., T. M. Gallie, Jr., M. J. Hodel, R. E. Hodel, F. J. Murray, D. A. Smith, and D. A. Smith, II. **MATHPROGRAM: A Computer Supplement for Calculus**. Iowa City: CONDUIT, 1980. (Apple II version, 1982.)

4. Smith, David A. **Interface: Calculus and the Computer, 2nd ed.** Philadelphia: Saunders College Publishing, 1984.

Computer Algebra Systems

John Hosack

Computer Algebra Systems (CAS's) with the capability of carrying out many of the operations of calculus, linear algebra, and differential equations are becoming widely available in a form suitable for use in undergraduate instruction. These systems, which allow the integrated use of symbolic, numerical, and graphical analysis, can substantially alter mathematics instruction. I discuss here the use of computer algebra systems in the undergraduate curriculum, especially in calculus instruction.

What are CAS's?

Computer algebra systems are computer packages that do *symbolic* as well as numerical mathematics. Computers have long been used for numerical computations such as evaluation of a polynomial for a given value of the variable. However, factoring a polynomial (possibly with symbolic coefficients), symbolic differentiation, and finding an antiderivative all require a symbolic system. CAS's are interactive: A user enters a command at the keyboard, and the answer is displayed on the screen. For example, to find the factors of

$$x + cx^2 - bx^2 - ax^2 - acx - bcx + abx + abc,$$

the user enters the command

factor(*x*^3 + *c***x*^2 – *b***x*^2 – *a***x*^2 – *a***c***x* – *b***c***x* + *a***b***x* + *a***b***c*)

and the answer appears:

$$(x - a)(x - b)(x + c).$$

To integrate $f(x) = \log(x^2 + a^2)$ with respect to x, the user defines the expression

f := log(*x*^2 + *a*^2)

and applies the integration operator

integrate(*f*, *x*)

The output is an indefinite integral:

$$-2x + 2a \arctan(\frac{x}{a}) + x \log(a^2 + x^2)$$

To obtain the sixth degree Taylor polynomial for f expanded about $x = 0$, the user enters

taylor(*f*, *x*, 0, 6)

and the computer responds with

$$2 \log(a) + \left[\frac{x^2}{a^2}\right] - \left[\frac{x^4}{2a^4}\right] + \left[\frac{x^6}{3a^6}\right].$$

The standard operations of a CAS include arbitrary precision arithmetic, algebraic simplification, calculus (differentiation, integration, power series), matrix algebra, and solutions of systems of equations and of differential equations. There are also utility programs for expression manipulation, such as extraction of parts of expressions. Some CAS's also allow numerical procedures, such as numerical integration or graphing. They all include high–level programming capabilities allowing user–written procedures as extensions to the system. On the other hand, using a CAS does not require any programming by the user. These systems are readily available to the "computer novice" with a 10 to 15 minute introduction, starting with how to turn on the computer.

Early CAS's ran only on mainframe computers and had limited distributions. The most widely available system, REDUCE [4], has been used in both Europe and the United States. Since the late 1970's some of the large systems (MACSYMA [9] and REDUCE) were ported to minicomputers, a new large system (SMP [13]) and a medium sized system (Maple [2]) were developed for minicomputers, and a small system (muMath [12]) was developed for microcomputers. (Maple has recently been ported to the Macintosh Plus.) The descriptions of systems as "large," "medium," and "small" refer to the amount of computer system resources required and to the range of problems handled. On the horizon are computer algebra systems, such as SCRATCHPAD II [7], that have built–in knowledge of more abstract algebraic structures, such as groups, rings, and modules. Some general descriptions of CAS's can be found in [8] and [14]. A review of computer algebra systems for educational use can be found in [5].

The current state of calculus.

Calculus courses serve as the standard entry point into the mathematics major and as service courses. Do they do a good job in either capacity? The primary focus of these courses, as shown by questions on examinations, is on carrying out algorithms for differentiation and integration. Even if students master the algorithms well enough to pass the course, they often do not remember the algorithms or are unable to apply them effectively in later work. Since most of the algorithms of calculus can be carried out by computer algebra systems, it seems inefficient to focus on the student's ability to do algorithms. Instead, computer algebra systems could be used to help change the emphasis from carrying out algorithms to conceptual understanding and problem solving.

There have been efforts in the past to use computers as tools in teaching calculus. These efforts have focused on numerical methods and graphics through both supplements and calculator/computer additions to standard texts, but they have not changed the way most calculus courses are taught. One reason is that most of these attempts require students to implement algorithms on computers or calculators. The typical calculus course is already overloaded with material, and few instructors believe that the pedagogical advantages of implementing numerical algorithms are worth the time, especially when the emphasis of the course remains on symbolic algorithms. Another factor is that much of calculus is concerned with symbolic operations, which are not suited to floating point numerical methods. Stoutemyer [11] points out that the arithmetic computations done with standard computer languages are not relevant to major portions of the mathematical curriculum (such as transformations of symbolic expressions, theorem proving, general solution of differential equations, and number theory), and even the numerical work encounters difficulties, since floating point arithmetic sometimes does not agree with the precise arithmetic done in school and available on computer algebra systems.

Computer algebra systems have a much greater opportunity to affect instruction than the previous efforts for two reasons. First, these systems are easy to use. They require no programming effort by the student and thus no class time for programming instruction. Second, the computer algebra systems perform the symbolic operations of calculus. Many of the CAS's have graphics and numerical analysis routines built–in; if they are not, then the desired algorithms can be added by the instructor, or the CAS can be interfaced to existing routines. Thus CAS's allow a unified approach to analysis using symbolic, numerical, and graphical methods.

Instructional advantages of CAS's.

It would be a mistake to incorporate CAS's into our courses primarily as exercise solvers while continuing our present emphasis on carrying out algorithmic computations by hand. Our goals, expectations, assignments, and classroom instruction need to change to maximize the opportunities afforded by modern technology. We briefly describe six areas in which a CAS can effect major changes in the undergraduate program.

(1) Student perception of what is important in mathematics.

Students normally measure the importance of an activity by the amount of time spent on it and the proportion of the examinations allotted to it. Since most of a student's effort on both homework and tests is devoted to algorithmic computation, it is not surprising that students view mathematics as a collection of formulas to be memorized, and "to do mathematics" is to compute. ("I can do the math, it's just the theorem I don't understand.") What does it mean when a student says "I understand integration?" Is the student referring to the convergence of Riemann Sums or the relationship between an antiderivative of a function and the area under its graph? Or does this mean that, given a list of ten integrals to evaluate in closed form, he or she can evaluate nine correctly?

Several things happen when routine algorithmic computations are relegated to a machine. Most important of all, time is made available for concentrating on concepts, motivation, applications, and ramifications. Computational mistakes are eliminated. (Think of the frustration and wasted time that result from a simple arithmetic mistake in reducing a matrix.) Self–confidence develops from understanding and applying concepts rather than from being a computational robot. Computation is viewed as a means rather than as an end. A CAS is an effective tool for convincing students that the appropriate focus should be on concepts and processes rather than on mastery of algorithms.

(2) The role of approximation and error bound analysis.

Approximation and error bound analysis are at the heart of applied mathematics as well as being the backbone of analysis. However, this analysis is only lightly touched on in elementary calculus because of the extensive computations that are often involved. Students are seldom asked to approximate an integral by Simpson's rule and determine an error bound because the computations are often too tedious and error prone. Thus, in this situation, the algebra is the restricting influence on both the type and level of inquiry. In a more general sense, algebraic limitations are a prime cause for the dependence of our elementary courses on closed form solutions. This in turn eliminates or restricts modeling in elementary courses to artificially constructed examples and leads students to question the value or applicability of what they are doing. A CAS is an effective tool for shifting the emphasis from closed form solutions to open ended problems. In particular, numerical integration can be developed as the "norm" or "standard," with closed form integration being considered as a special case. The availability of CAS's can open the door to modeling in our introductory classes, so that our students can experience more realistic applications.

(3) Integration of problem solving methods.

Currently students learn to apply a set of symbolic algorithms to solve problems. For example, to integrate a rational function, the student learns the method of partial fractions. However, this method can be applied only when the denominator can be factored. Since it is generally impossible to find the roots of polynomials by symbolic methods, this method is applicable only in special cases. Computer algebra systems allow the student to gain experience in combined use of symbolic, numerical approximation, and graphical methods for solving problems. An example of this integrated approach is given in the next section.

(4) Development of problem solving skills.

To many students there are only two problem solving skills: finding a suitable worked example to mimic, and carrying out algorithmic computation. This is a consequence of our computationally based instruction, with its emphasis on facts and techniques. To be an effective problem solver, one needs to consider alternatives, to experiment, to conjecture and test, and to analyze the results. Attitude, therefore, is a major aspect of problem solving. Shifting the burden of computation to a CAS makes time available for students to concentrate on how to approach a problem, delineate subproblems, consider alternatives, and experiment.

(5) Development of an exploratory approach to learning mathematics.

Students do not normally regard themselves as active participants in mathematical exploration, but rather as passive recipients of a fixed body of facts and algorithms, another consequence of the limitations imposed by hand computations. Students spend considerable time on convergence tests for series, but the question of the sum of a convergent series (even in approximate form) is seldom considered. Likewise, questions about the rate of convergence are usually not raised. Our analysis is "one level deep," in the sense that we do not encourage students to pursue follow–up questions. In fact, these questions are seldom even raised in our classrooms. Computer algebra systems encourage students to develop an exploratory approach to learning by making computations almost effortless. Computer aided analysis becomes a standard process for investigating and developing conjectures.

(6) Exercises and tests.

Most homework exercises and test problems are designed for developing and measuring skill in carrying out algorithmic processes that can be done faster and more accurately by computer algebra systems. We should use CAS's as tools that allow us to challenge our students with conceptual understanding, problem solving, and exploration. Here are some possible uses of a CAS in exercises:

(a) "Open ended" problems, such as determining the contour of a terrace subject to certain boundary conditions.

(b) Approximating integrals or sums of series within stated error bounds or determining error bounds for such approximations.

(c) Making conjectures and then proving or disproving them. (For example, what are the geometrical interpretations of the multiplicities of the zeros of numerator and denominator of a rational function?) In a structured sequence of exercises with a CAS, a student can be led to examine enough cases to generate a conjecture.

(d) Construction of examples that satisfy certain constraints. Exercises in which students make up examples are an effective way to enhance conceptual understanding. The CAS allows students to easily examine many cases and test conjectures, which helps them to either find the desired example or show why no such example can exist.

(e) Problems drawn from realistic models that are not feasible to do by hand computation.

(f) Student–invented problems for which they can use the CAS to check their answers.

An example of the use of a CAS.

In this section we apply Simpson's Rule to a problem with a unified set of numerical, graphical and symbolic tools provided by a CAS.

Suppose we want to compute $\int_1^4 \sin x^2\, dx$ with an error of at most 10^{-4}. The first thing we might try is to use our CAS to find an antiderivative:

integrate(sin(x^2),x).

Our CAS will either fail, or, if it is sophisticated, may give the answer as an expression involving complex numbers and the error function, erf.

Next we try a numerical approximation, in this case Simpson's Rule. First we need to determine the number of intervals, $2n$, required to obtain the desired accuracy. That is, we want to use the error bound to find an n that satisfies

$$\frac{B(b-a)^5}{2880n^4} \le 10^{-4},$$

where $f(x) = \sin x^2$, and B is a bound for $|f^{(4)}|$ on [1,4]. We can compute $f^{(4)}$ using the symbolic capabilities of our CAS:

d4f := diff(sin(x^2),x,4)

This gives:

$$\text{d4f} = 16\sin(x^2)\,x^4 - 48\cos(x^2)\,x^2 - 12\sin(x^2)$$

We need to estimate the maximum of d4f on [1,4]. We can try to do this by using calculus. We know that, since d4f is a differentiable function, its derivative

$$\text{d5f} = 32\cos(x^2)\,x^5 + 160\sin(x^2)\,x^3 - 120\cos(x^2)\,x$$

will vanish at a point in (1,4) where d4f has a maximum. If we ask our CAS to

solve(d5f=0,x)

we may get several solutions or no solutions, since "solve" is designed to find exact, symbolic solutions rather than approximate numeric solutions. A good idea would be to look at the graph of d4f to see what to expect. We may be able to obtain a sufficiently good estimate of the bound B from the graph alone:

graph(d4f,[1,4])

gives

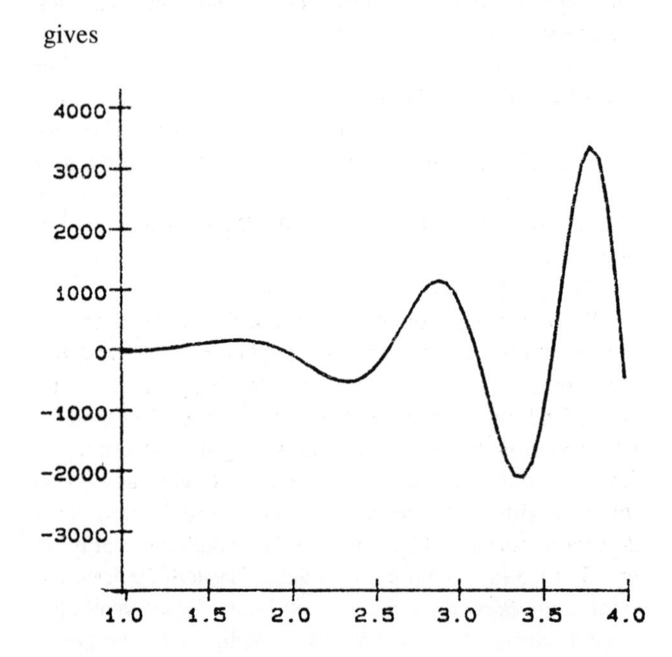

We see that d4f has several local extrema in (1,4), that $|f^{(4)}|$ attains its maximum on [1,4] at a point between x =3.5 and x = 4, and that the maximum appears to be less than 3500. If we need a more precise answer, we can use the bisection algorithm to solve for a zero of d5f in the interval [3.5,4] with an error of at most 10^{-6}:

bisect(d5f,x,[3.5,4],10^(–6))

returns 3.804649831. The value of d4f at this point is $f^{(4)}(3.8046498) \approx 3382$. Thus we have found a bound B for $|f^{(4)}|$ on [1,4] of 3500, or, more precisely, 3382.

We can now compute the number of intervals required to guarantee the desired accuracy of 0.0001. We need to solve

$$\frac{3382(4\ -\ 1)^5}{2880\ n^4} \le 10^{-4},$$

or

$$n^4 \ge \frac{10^4 \cdot 3382 \cdot 3^5}{2880}.$$

We can use the CAS as a calculator to evaluate

$$\left[\frac{10^4 \cdot 3382 \cdot 3^5}{2880}\right]^{1/4} \approx 41.1.$$

Thus n should be at least 42, so we will use 84 subintervals.

We can now apply Simpson's Rule:

simpson(sin(x^2),x,[1,4],84)

returns the value 0.436870, or 0.4369 to four significant digits.

The number of subintervals used, 84, is based not on the specific function, $f(x) = \sin x^2$, but on the bound for the fourth derivative, so this number is guaranteed to work for any function with the same bound. Thus the number of subintervals is likely to be conservative; fewer subintervals may give the desired accuracy. For this particular function, we can experiment and determine that for 42 or more subintervals the value of the Simpson's Rule approximation is 0.4369 to four significant digits.

Current use of CAS's.

The use of computer algebra systems in undergraduate instruction is currently at an exploratory stage. Several recent experimental efforts are briefly described here.

At Colby College experimental use of CAS's (primarily Maple) has been made in calculus and some upper division courses. The object has been to identify possible uses, problems, and areas for future work [6], [10].

At the University of Waterloo, Maple was made available to many sections of undergraduate mathematics classes [1]. Some problem solving exercises were provided as options for calculus students. Students in upper division courses have also used Maple on an optional basis.

St. Olaf College has experimented with an SMP–based applied calculus course [15]. The emphasis was on integration of discrete concepts into calculus (approximation, numerical and constructive techniques), with SMP used to handle the routine computations and graphing.

At the University of Hawaii, muMath has been used as an aid to classroom instruction [3]. The standard muMath capabilities were enhanced with graphics capabilities. muMath has also been used to supplement calculus and linear algebra courses at Gettysburg College.

What is needed for effective use of a CAS.

Computer algebra systems, as initially developed, were oriented towards the individual researcher, with little thought given to educational use. The newer systems, such as Maple and SMP, have more of the needed capabilities. In fact, instructional use was an explicit goal in the design of Maple. This section discusses some of the hardware and software capabilities needed for computer algebra systems. Reference [5] compares some of the available CAS's as instructional aids.

The system must be easy to use. Most instructors cannot devote a significant amount of class time to teaching the mechanics of the use of these systems. Happily, most of the existing systems have easily remembered instructions for common tasks. Most systems (e.g., SMP, Maple, MACSYMA) have on–line help, which is very useful if the manual is not handy. Many questions about usage can be answered if the on–line help provides a few simple examples. Allowing intermediate steps to be shown is a useful option; MACSYMA has limited capabilities for such "verbose" output.

The CAS should have well integrated facilities for symbolic, numerical, and graphical analysis. Most of the desired operations should be built–in and not just potentially definable. These systems allow the user to extend the built–in operations by defining new operations as user defined subroutines. Because of constraints on time, it is probably not feasible for most students to develop the skills required to define any but the simplest subroutines. However, there should be a good language for defining new operations for use by the instructor and advanced students. User developed algorithms should be compilable into machine language to permit efficient, rapid execution. It is also desirable to be able to interface the CAS with existing numerical packages, allowing a symbolic expression to be passed to the numerical package.

The output should be in a form that is easy to read and interpret. Some systems have only one–dimensional output, i.e., the two–dimensional expression

$$\frac{(c - d)^2}{a - b}$$

is printed as

$$(c - d)**2/(a - b)$$

One–dimensional output can make it difficult to read long expressions.

The results of the computations should be in a form accessible to the student. For example, in an earlier version of SMP the standard integration routine gave the integral of a trigonometric function as a complex exponential; it had a separate operation to transform the answer to trigonometric form. In the current version this transformation is done by default.

If a time–sharing system is used, the CAS should not overburden the computer system. Most computer algebra systems require substantial computer resources. Thus it may be impossible for more than a few people to use the CAS at one time. Experiments combining CAS with graphics [1], [5] suggest that graphics places too much of a burden on a time–sharing system for use by large classes. The best environment for a CAS is a powerful microcomputer, an "academic workstation." These are becoming available, but the cost is still high. An important step was made in early 1987 when Maple became available for the Macintosh Plus.

Some problems and questions.

(1) Is the necessary hardware available? The current hardware situation is not ideal, but as CAS's are ported to workstations and the prices of powerful workstations drop in the next few years, the hardware problems will be resolved.

(2) Will students using a CAS tend to "compute" before thinking? This is a real danger *unless* we change our instruction, assignments, and objectives to use CAS's for the enhancement of learning. Using a CAS for a homework assignment to differentiate or integrate a collection of functions from a standard calculus text would be a waste of time, yield negative learning, and lead to the "button pushing robot" syndrome. On the other hand, using a CAS to find the fourth or fifth derivative to determine the error bound in an application of Simpson's Rule, or to obtain and sketch partial sum approximations to illustrate series convergence, or to do needed computations that are not tractable by hand, would yield positive learning. The value of a CAS as an educational tool is dependent on the types and objectives of the tasks for which it is used.

(3) How will the widespread use of CAS's affect the beneficial aspects of hand computation? Many students have been attracted to mathematics because of their enjoyment of and success with computations. How will the de–emphasis of hand computation affect these students? Perhaps future students will be attracted to mathematics because of their success in the secondary schools with applying computers to problems. Some students have gained mathematical self–confidence from successfully completing routine computations; in the future, this confidence will be a result of completing a multi–faceted analysis of a problem with the aid of a computer.

(4) Will using a CAS erode important computational skills? The importance attached to a computational skill is relative rather than inherent; we no longer teach methods of extracting roots of integers by hand computation. Thus, changing technology and changing needs are constantly altering the importance attached to any particular computational skill. However, using mathematics to describe and understand our world is and will remain of prime importance, independent of any particular mode of computation. CAS's will effect changes in computational skills. For example, we expect a decrease in students' abilities to recall and carry out many standard algorithms, such as those for factoring and integration. These skills are seldom remembered well and will be of little importance in future environments where CAS's will be common. We believe that the gain in conceptual understanding and problem solving skills is well worth the trade off.

(5) How should the curriculum be changed to reflect and take advantage of the widespread availability of CAS's? The opportunities presented by CAS's require a significant modification of existing mathematical curricula. What traditional computational methods should be retained? How should their motivation and treatment be changed? What mixture of closed–form, series, and approximate numerical methods should be used in calculus? More time will have to be spent on deciding which approach to solving a problem is best, rather than routinely considering only one technique. Restructuring the curriculum appears to be the most difficult problem in using a CAS. Texts and other instructional material will have to be created. Since CAS's will become accessible to students (whether part of a mathematics program or otherwise), such changes in curricula will become essential.

I thank Don Small, Bruce Char, Paul Zorn, and Carl Leinbach for their suggestions.

REFERENCES

1. Char, B. W., K. O. Geddes, G. H. Gonnet, B. J. Marshman, and P. J. Ponzo. "Computer Algebra in the Undergraduate Mathematics Classroom." **Proceedings of the 1986 Symposium on Symbolic and Algebraic Computation**. New York: The Association for Computing Machinery, 1986.

2. Char, B. W., K. O. Geddes, G. H. Gonnet, and S. M. Watt. **Maple User's Guide**. Waterloo, Ontario: Watcom Publications, 1985.

3. Freese, R., P. Lounesto, and D. A. Stegenga. "The Use of MuMath in the Calculus Classroom." To appear in **The Journal of Computers in Mathematics and Science Teaching**.

4. Hearn, A. **REDUCE User's Manual**. The Rand Corporation, 1983.

5. Hosack, J. M. "A Guide to Computer Algebra Systems." **The College Mathematics Journal 17** (1986), 434–441.

6. Hosack, J. M., K. Lane, and D. B. Small. "Report on the Use of a Symbolic Mathematics System in Undergraduate Instruction." **SIGSAM Bulletin 19,** 1 (February, 1985), 19–22.

7. Jenks, R. D. "A Primer: 11 Keys to New SCRATCHPAD." **EUROSAM 84**, J. Fitch (editor). New York: Springer–Verlag, 1984.

8. Norman, A. C. "Algebraic Manipulation." **Encyclopedia of Computer Science and Engineering**. New York: Van Nostrand Reinhold, 1983, pp. 41–50.

9. Pavelle, R., M. Rothstein, and J. Fitch. "Computer Algebra." **Scientific American 245**, 6 (December, 1981), p. 145.

10. Small, D. B., J. M. Hosack, and K. Lane. "Computer Algebra Systems in Undergraduate Instruction." **The College Mathematics Journal 17** (1986), 423–433.

11. Stoutemyer, D. R. "A Radical Proposal for Computer Algebra in Education." **SIGSAM Bulletin 19,** 1 (February, 1985), 40–53.

12. Wilf, H. S. "The Disk With the College Education." **American Mathematical Monthly 89** (1982), 4–8.

13. Wolfram, S. "Computer Software in Science and Mathematics." **Scientific American 251**, 3 (September, 1984), 188–203.

14. Yun, D. Y. Y., and D. R. Stoutemyer, "Symbolic Mathematical Computation," **Encyclopedia of Computer Science and Technology**, vol. 15, pp. 235–310. New York: M. Dekker, 1980.

15. Zorn, P. "Calculus from a Discrete Viewpoint: A Course with Computer Symbolic Manipulation." Presented at the MAA Meeting in Laramie, Wyoming, August, 1985.

The Use of Computers in Teaching Discrete Mathematics

Martha J. Siegel

Humorist Fran Liebowitz, giving "Tips for Teens" [4], advises "Stand firm in your refusal to remain conscious during algebra. In real life, I assure you, there is no such thing as algebra." Mathematics teachers are frequently faced with semi-conscious students, many of whom are convinced that, indeed, there is no such thing as algebra. Enter technology: computers, video screens, sound effects! Now we can get the attention of our audience! But are we teaching mathematics?

Recent research has shown that doing mathematics requires both hemispheres of the brain. Successful mathematicians use the symbolic, conceptual, linguistic, and analytic left brain in concert with the spatial, non–verbal, nonsymbolic right brain. The left brain is associated with number skills, written and spoken language, reasoning and scientific skills. But the right brain controls insight and imagination as well as spatial ability [9]. When teaching, we try to encourage students to use the whole brain by example: being verbal, writing proofs, drawing pictures, trying to put results in a holistic setting.

Can the computer help us to teach mathematics more effectively? Can we use computers to gain more than attention? Can we use them to help students to integrate a variety of modalities in attacking a problem? In this article I shall address these questions in the area of discrete mathematics, but many of the rewards and pitfalls are common to other areas as well.

Courses in Discrete Mathematics.

Discrete mathematics is the stuff computers do! It is the mathematics of finite and countable sets, and it includes topics taught throughout the standard secondary and college curricula. These topics include logic, set theory, combinatorics, discrete probability, functions and relations on discrete structures, induction, recursion, difference equations, graph theory, trees, algebraic structures, and linear algebra.

Many of the topics are scattered through the secondary school curriculum — more so in the 1960's than in the "back to basics" 1970's. But in the mid–1980's, they are making an appearance again. Their relevance to computer science is the motivation for the recent attention. The modern emphases are on algorithms and on the relationship of these topics to computer science. In many colleges and universities, courses at the junior–senior level devoted exclusively to parts of this material have been taught for at least ten years. For example, there are many courses and texts in discrete structures and applied algebra (typical examples are [5] and [2]). Courses of comparable sophistication in combinatorics or graph theory have also been featured as electives in many schools. A fuller description of these courses can be found in the 1981 CUPM Report [8].

More recently, one– and two–semester courses at the freshman–sophomore level have been introduced. These are on the level of the calculus, with a prerequisite of either four years of high school mathematics or elementary calculus. These courses often are required for both mathematics and computer science majors and are taught as complementary to the first computer science course; they also serve as prerequisite to higher level courses in both disciplines. The *Final Report of the MAA Committee on Discrete Mathematics in the First Two Years* [7], available from the MAA, fully outlines the nature of these courses. The report also contains an extensive annotated bibliography of the many textbooks published recently.

What are our pedagogical goals in teaching discrete mathematics? At the elementary level, we want to introduce a student to a way of thinking, a way of bringing some order to the chaos by way of mathematical structures. We hope that students will gain an understanding of the propositional calculus, become adept at handling quantified statements, be able to do simple proofs. We hope to see students write correct proofs by induction. We teach students to construct recursive relations to solve a problem, and we help them develop effective algorithms and prove their correctness. One might say that we are striving to help students become mathematically mature.

Can the computer help us to attain such lofty goals? Ideally, the relationship should be symbiotic; that is, the use of the computer should enhance the understanding of the mathematics and the mathematics should enhance the understanding of computers. The differences in cognitive style required by the potpourri of topics gathered together as discrete mathematics pose a problem. Are there essential differences between algorithmic thinking and mathematical thinking? Donald Knuth discusses this question in a recent article [3] and answers "yes". And yet, the ideal is that students take the first computer science course and elementary discrete mathematics concurrently. Then conscientious faculty cooperating in the teaching of the two courses can create for the student an atmosphere in which algebra not only "exists" but is useful. Different modes of thinking can be applied consciously, and the computer can be a significant aid.

Software for Teaching Discrete Mathematics.

As the computer is so useful for the illustration of algorithms, both logically and graphically, it is disturbing that I cannot report that there is a significant amount of software commercially available for the teaching of discrete mathematics at any level. I am omitting any mention of the many packages available for linear algebra or for linear programming, some of which are discussed in the papers by Orzech and by Herman, et al., in this volume. There are two packages written specifically for use in discrete mathematics. They are very different from one another, and each offers the teacher an opportunity for experimentation.

Accompanying Marvin Marcus' *Discrete Mathematics* [6] is a disk containing all of the BASIC programs and routines contained in the appendix, the computer exercises, the narrative, and the problem sections. The disk is available for either the IBM PC or the Apple II+ and IIe. The contents allow the reader (user) to experiment with and change the many algorithms presented in the text without having to reenter the programs. A knowledge of BASIC is desirable, but the text has the essentials for the novice. Exercises in the text lead the student to the development of programs which illustrate relations, functions, and combinatorics in an innovative way. They encourage trial–and–error and conjecture. Exercises in the book also demand proofs.

In each of the ten chapters there are computer exercises as well as short answer quizzes and pencil–and–paper exercises for which the computer examples provide insight. The chapter headings correspond to the topics covered by the computer package: Elementary Logic, Sets, Relations and Functions, Some Important Functions (exponential, logarithmic, and polynomial; permutations and sorting; string functions), Function Optimization, Induction and Combinatorics, Probability, Matrices, Linear Equations, and Linear Programming. Applications of the Monte Carlo method abound. Even if one did not use the text, many of the programs are excellent additions to one's algorithmic treasure chest and can be profitably introduced to classes at all levels, especially with Marcus' good questions about them.

The other major software package for discrete mathematics is the recently released *Discrete Mathematics* by John G. Kemeny, a utility package that can be used with students in various ways. The programs are user–friendly, require no knowledge of programming, and can be used on an IBM PC or compatible, an Apple Macintosh, or a Commodore Amiga (see Software Sources at the end of the article). There are nine routines: *truth tables*, *Venn diagrams*, *counting*, *recursive*, *graphs*, *binary*, *sorting*, *equations*, and *algorithms*. The manual is easy to read, and the graphics are excellent. The user with access to True BASIC can modify the programs, making the package far more valuable as a learning tool. A brief description of the routines will give the flavor of the package.

> *Truth tables* generates truth tables for logical statements entered by the user with up to four variables and a number of logical connectives (and, not, or, exclusive or, if...then, if and only if). One can enter a single proposition and get its truth table, or one can enter two propositions to get a truth table for each and a statement as to whether one logically implies the other.
>
> *Venn diagrams*: The user enters up to three set names and can request the Venn diagrams of intersections, unions, set differences, complements or the universal set. Another option is to enter the number of elements (or the probability) in each of the sets. Then the user can request that the Venn diagram show the corresponding number of elements (or probability) in some combination of the sets.
>
> *Counting* contains routines for calculating combinations, permutations, partitions (the number of ways of dividing n things into cells of specified sizes), boxes (the number of ways of arranging n identical objects in j boxes), and the Pascal triangle.
>
> *Recursive* allows the user to define a function recursively and request the value $f(n)$ for any natural number n for which the function is properly defined.

Graphs contains options for drawing the graphs and digraphs (weighted or unweighted) specified by the user's input of vertices and edges. One can also request the minimal path from ... to ...; check to see if there are Euler or Hamiltonian paths and circuits; or check if there is a spanning tree from a given vertex.

Binary provides a graphic demonstration of a binary search and a binary tree for the user's list of words.

Sorting demonstrates bubble sort, selection sort, sort-merge, and quicksort for a random list of numbers. A summary of the number of comparisons, number of swaps, and the elapsed time used by each sort is supplied also.

Equations solves systems of linear equations with up to eight unknowns.

Algorithms lists the source code for several of the algorithms mentioned earlier. If you have access to True BASIC, these algorithms can be modified.

An innovative teacher can use Kemeny's *Discrete Mathematics* to help students understand the material in an elementary course. Unlike the Marcus disk, there are no questions provided that might lead the student to use the programs as a learning tool, so the teacher must consider the text chosen for the course and provide appropriate supplementary material to tie all of the elements of the course together. If the course were oriented toward discrete models, then the utility aspect of the disk would be exceptionally useful. Without incisive exercises from the instructor, most students are likely to be impressed with the graphics and speed and then say, "So what?".

Many of the new packages written for student use in operations research can also be used effectively in a discrete mathematics classroom. One such package is *Microsolve/Operations Research*, which contains routines for network flow programming. These models are, of course, the graph–theoretic models of the discrete mathematics course, and the minimum path algorithms are part of the optimization goal. The programs are written in compiled BASIC and run on an IBM PC with at least 128K and DOS 2.0 or higher. The package includes Markov chains, simulation of queues, and birth and death processes, any of which could also enhance the discrete mathematics course. There are other operations research packages that contain similar programs, some of which might be useful for the discrete mathematics course.

In addition to software written specifically for use in particular courses, there are the powerful symbolic computational programs, such as Maple, which can be used in a creative way in the classroom. There are several of these computer algebra packages on the market, most of them particularly useful to the applied mathematician and researcher. Maple has some features which might make it the package of choice of this type for discrete mathematics instruction. For example, set theory operations and number theory routines are featured. The package was developed for ease of student use, and students at the University of Waterloo have been using it in calculus and in applied mathematics courses which include combinatorics. The development of instructional materials to fully integrate the available computer power into the syllabus in a meaningful way is still in process. The symbolic algebra system deserves the attention of creative mathematics instructors, for its capabilities in the classroom have yet to be exploited. (See the article by John Hosack in this volume for further details.)

Conclusion.

The computer offers an exciting opportunity for enhancing the learning of discrete mathematics. To be successful, the computer materials must be integrated with the textbook, must incorporate the elements of proof, and must allow for trial and error, analysis, and creativity in the development of algorithms and models. We have much to learn about this task. However, the current research in artificial intelligence and knowledge–based systems indicates that computing is a powerful tool which improves our ability to help students integrate the symbolic, the graphic, the analytic, and the creative in their learning of mathematics.

REFERENCES

1. Char, D. W., K. O. Geddes and G. H. Gonnet. "The Maple symbolic computational system." **ACM SIGSAM Bulletin 17**,3 (1983), 31–42.

2. Dornhoff, Larry L., and Franz E. Hoff. **Applied Modern Algebra**. New York: Macmillan Publishing Co., 1978.

3. Knuth, Donald E. "Algorithmic thinking and mathematical thinking." **The American Mathematical Monthly 92** (1985), 170–181.

4. Liebowitz, Fran. **Social Studies**. New York: Random House, 1981.

5. Liu, C. L. **Elements of Discrete Mathematics**, 2nd ed. New York: McGraw–Hill, 1985.

6. Marcus, Marvin. **Discrete Mathematics: A Computational Approach Using BASIC**. Rockville, MD: Computer Science Press, 1983.

7. Mathematical Association of America. **Final Report of the MAA Committee on Discrete Mathematics in the First Two Years**. Washington, DC: MAA, 1986.

8. Mathematical Association of America. **Recommendations for a General Mathematical Sciences Program**, Report of the Committee on the Undergraduate Program in Mathematics. Washington, DC: MAA, 1981.

9. Restak, Richard. **The Brain**. New York: Bantam Books, 1984.

SOFTWARE SOURCES

Microsolve/Operations Research, IBM Revised Edition. Paul A. Jensen. Available from Holden–Day, Inc., 1985.

Discrete Mathematics: A Computational Approach Using BASIC. Marvin Marcus. Available from Computer Science Press, 1985; for Apple II+ and IIe or IBM PC.

Discrete Mathematics. John G. Kemeny. Available from True BASIC, Inc., 1986; for IBM PC with 256K and IBM graphics adaptor, IBM–compatible MS–DOS computers, 512K Apple Macintosh, and 512K Commodore Amiga.

Maple. Symbolic Computation Group, University of Waterloo, Waterloo, Ontario, Canada. Available for the VAX and MicroVAX from WATCOM Products, Inc., 415 Phillip Street, Waterloo, Ontario, Canada N2L 3X2 (519–886–3700).

Finite Differences: A Computer – Based Alternative to Calculus

James T. Sandefur and Andrew Vogt

The computer age has given new impetus to an old branch of mathematics, the method of finite differences, which treats problems of time evolution posed in discrete rather than continuous form. A discrete mathematics centered on difference equations is timely, not only because of the rise of computers, but because of increasingly evident shortcomings, both pedagogical and conceptual, in the calculus. Discrete mathematics also creates a link between ideas and techniques of computer science and important and useful mathematical notions. An elementary course on the subject can even bring beginning students into contact with active research in mathematics, physics, chemistry, and other areas.

The calculus of Newton and Leibniz was designed to circumvent the difficulties of dealing with the discrete by passage to continuous limits. Sums become integrals, differences become differentials, and the computational labor of repeated additions and subtractions is swept away by the power of the infinitesimal calculus.

However, this great tool has shortcomings. First, it is not easily understood. Many otherwise intelligent people turn away from science because they cannot make the step from a high school mathematics dwelling on the finite and discrete to a higher mathematics of the infinite and continuous. The abrupt and often poorly motivated transition to the difficult concepts of calculus has altered the career path of many a would–be engineer, physicist, chemist, or economist.

Another difficulty with traditional calculus is that many modern problems resist continuous methods. Even the student who succeeds with calculus needs to appreciate discrete approximation schemes which can be implemented on the computer. Such schemes are increasingly important for the solution of the analytically intractable differential equations arising in many applications. In the real world, integrals are replaced by sums and differentials by differences, instead of vice versa!

A course in finite difference methods offers hope of remedying these ills. It will appeal to talented students who might go elsewhere because of the unimaginative syllabus of the typical calculus course (see the thoughtful critique of Ungar [8]). It will also better prepare students for the coming realignment in importance of the continuous and the discrete, made possible by the computer and necessitated by the mathematical complexity of modern scientific problems.

In such a course connections between mathematics and computer science arise naturally. A calculator or computer is an obvious aid for numerical work in discrete mathematics. Induction, a basic technique in the study of difference equations, can be linked with the programming concept of recursion, and the student's programming skills can be developed while he or she learns the mathematics. Computer graphics can be used to give the student geometric insight into perhaps the greatest idea to come out of the Newtonian calculus, that of an evolving dynamical system.

One of us (JTS) has taught a course of this type for several years, based on the notes [7]. This course is described in the remainder of this article through a series of examples, the order of presentation illustrating one way to develop the subject.

First–order linear difference equations.

Consider the first–order linear system

$$X(n+1) = r\,X(n) + b \text{ for } n \geq 0, \text{ with } X(0) = x_0, \quad (1)$$

where r, b, and x_0 are given real numbers. We think of $X(n)$ as the state of the system at time n. The first equation tells us how the state evolves from one time to the next; it is a *difference equation* that specifies how $X(n+1)$ differs from $X(n)$. The other equation gives the initial state. Our aim is to predict future states, that is, to find $X(n)$ for all $n \geq 1$.

To make matters more concrete, (1) may be regarded as a population growth model, with $X(n)$ the population at the beginning of year n, $r - 1$ the net annual growth rate of the population, and b the net annual immigration (number of immigrants minus number of emigrants). Another interpretation of (1) is as a savings model, with $X(n)$ the amount in the bank at the beginning of the n–th interest period, $r - 1$ the interest rate per period, and b a constant amount deposited or withdrawn at the end of each period.

For particular choices of r, b, and x_0, the student can easily compute $X(1)$, $X(2)$, ... by calculator or computer. In contrast with the abstract manipulations of calculus, this routine computation gives the student a feel for the dynamics of the system and an awareness of the state as an evolving entity. The interpretations heighten this awareness: The first model demonstrates the ideas of Thomas Malthus on population growth, while the second illustrates the benefits of thrift.

With computer graphics the student can explore the dependence of the evolution on the parameters r and b. In Figure 1 successive points $(n, X(n))$ are plotted for various choices of r with $b = 0$ and $x_0 = 4$. Exponential growth occurs when $r > 1$, steady state when $r = 1$ (but only for $b = 0$), exponential decay when $1 > r > 0$, and damped oscillation when $0 > r > -1$. For $r = 1$, $X(n)$ is a linear function of n, increasing if $b > 0$, decreasing if $b < 0$. For r different from 1, there is no qualitative dependence on b: Trajectories similar to those in Figure 1 occur with 0 on the vertical axis replaced by $b/(1-r)$. In Figure 2 a damped oscillation is shown with $x_0 = -2$, $b = 5.4$, and $r = -0.8$.

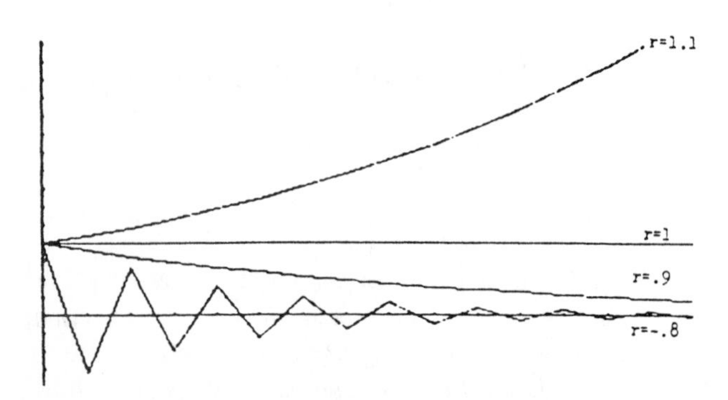

Figure 1. Trajectories of $X(n+1) = r\,X(n)$.

The state $b/(1-r)$ is an *equilibrium* state, that is, a constant solution. If $|r| < 1$, the system evolves toward this state (as in Figure 2), no matter what x_0 is. The equilibrium in this case is said to be *stable*: Departures from it (due perhaps to momentary disturbances of the system not included in the dynamical law) will die out as time passes. The stable equilibrium is also called an *attractor*: Other states evolve toward it.

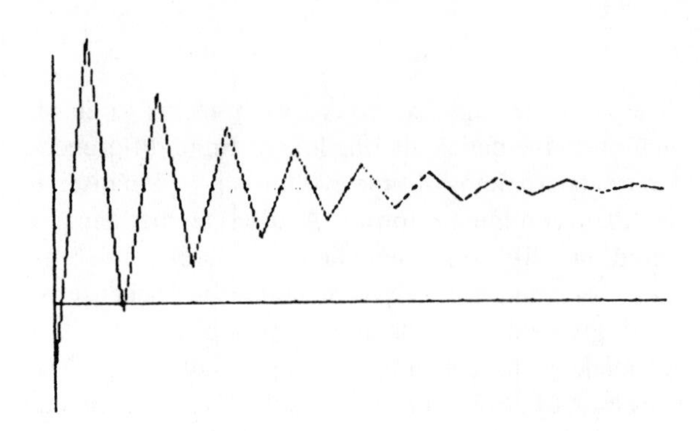

Figure 2. $X(n+1) = -0.8\,X(n) + 5.4$, $X(0) = -2$.

When $|r| > 1$, non–equilibrium states evolve away from $b/(1-r)$, which is then an *unstable* equilibrium. If $r = 1$, either every state is an equilibrium (when $b = 0$), or there are no equilibria ($b \neq 0$). If $r = -1$, the equilibrium state is $b/2$. Other states neither approach it nor flee it, so $b/2$ is said to be a *neutral* equilibrium. All other states lie on *two-cycles*, oscillating between x_0 and $b - x_0$.

A striking application of computer graphics is the *cobweb* graph. The lines $Y = X$ and $Y = rX + b$ are plotted in the X–Y plane. These lines intersect at the equilibrium value $X = b/(1-r)$. We superimpose on this graph a series of zigs and zags, a cobweb, arrived at as

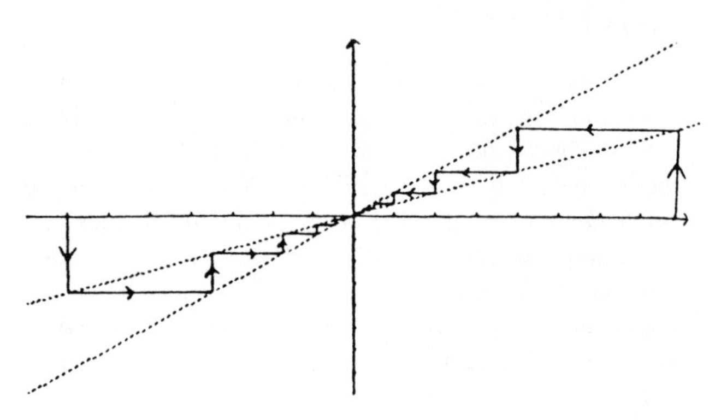

Figure 3. Cobweb graphs for $X(n+1) = 0.5\,X(n)$.

follows. Starting at the point $(X(0),0)$, we move vertically to $Y = rX + b$, arriving at the point $(X(0),X(1))$. Then we move horizontally to $Y = X$, arriving at the point $(X(1),X(1))$. We move vertically again to $Y = rX + b$ and $(X(1),X(2))$, then horizontally again to $Y = X$ and $(X(2),X(2))$. Continuing in this manner, we generate the sequence of points $\{(X(n),X(n))\colon n = 1,2,\ldots\}$ along the line $Y = X$. The two webs in Figure 3 (with $r = 0.5$, $b = 0$, and $x_0 = -7$ or $+8$) show the points $X(n)$ tending to 0 as n gets large. Similar graphs display exponential growth away from $b/(1-r)$ when $|r| > 1$, two–cycles when $r = -1$, and so forth.

Numerical and graphical work gives the student an appreciation for the analytical solution of (1). When $b = 0$, this solution is

$$X(n) = r^n x_0,$$

and the general case is a translation of this, namely,

$$X(n) = r^n(x_0 - c) + c, \qquad (2)$$

where $c = b/(1-r)$. This formula can be used to compute the monthly payment on a mortgage (with b as the unknown) or the amount of capital needed to fund a retirement annuity for some fixed number of years (x_0 unknown).

System (1) and formula (2) can also be used to prove the cobweb theorem of economics, that price is stable only if consumers are more sensitive to price than producers (see [7] and [4]). This can lead to interesting discussions about price supports and related matters.

Nonlinearity.

The computer plays an especially significant role when the difference equations under study are nonlinear. Entirely new mathematical phenomena have been discovered with the aid of the computer, and these phenomena continue to be explored by computer in the hope that clues to their nature will be discovered.

Consider the nonlinear "logistic" equation:

$$X(n+1) = rX(n)(1 - kX(n)) \text{ for } n \geq 0, \text{ with } X(0) = x_0. \qquad (3)$$

This models population growth under conditions of crowding, with $r - 1$ representing the net growth rate when the population is small and k the crowding factor, which comes into play as the population grows and competition develops for food and space. (If $k = 0$, the system reduces to (1) with $b = 0$.) If we attempt to investigate this system analytically, we immediately run into trouble. We would like to have a general formula for $X(n)$ in terms of r, k, x_0, and n, but $X(n)$ is a polynomial of degree 2^n in x_0 with coefficients of increasing complexity as n increases.

For convenience we assume $k = 1$, which is equivalent to a scale change in X. The equilibrium states of (3) are then easily seen to be $X = 0$ and $X = (r-1)/r$. To determine the stability of these equilibria, we *linearize* the equation. Expanding the nonlinearity about an equilibrium, and keeping the linear terms only, we obtain a linear difference equation with the same stability properties as the original equation (see [7] for details). We find that $X = 0$ is stable if $-1 < r < 1$ and that $X = (r-1)/r$ is stable if $1 < r < 3$. When $r = 1$, the two equilibria reduce to one at 0, which is *semistable* (stable from the right, unstable from the left). This is illustrated by the cobweb graphs in Figure 4, in which the line $Y = X$ and the curve $Y = X(1-X)$ are plotted. Although these two curves are tangent at $X = 0$, they do not cross there, and this implies semistability.

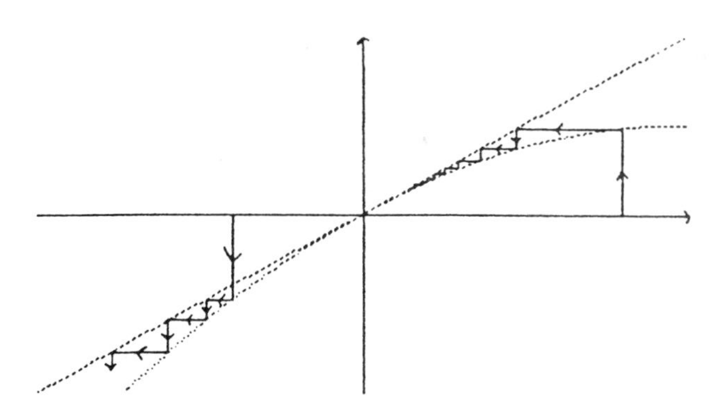

Figure 4. Cobweb graphs for $X(n+1) = X(n)(1 - X(n))$.

What happens when r is 3 or larger? Graphical investigation and theoretical follow–up show that cycles become important if r is in this region. For r slightly above 3, a two–cycle emerges. This emergence is called a *bifurcation*. The two–cycle is stable in the sense that a state near one of the two states of the cycle oscillates back and forth, tending to one state of the cycle for even n and to the other for odd n. This can be seen in the cobweb of Figure 5 ($r = 3.2$, $x_0 = 0.6$), in which $X(n)$, instead of tending to the equilibrium value, approaches the two–cycle represented by the rectangle in heavy lines.

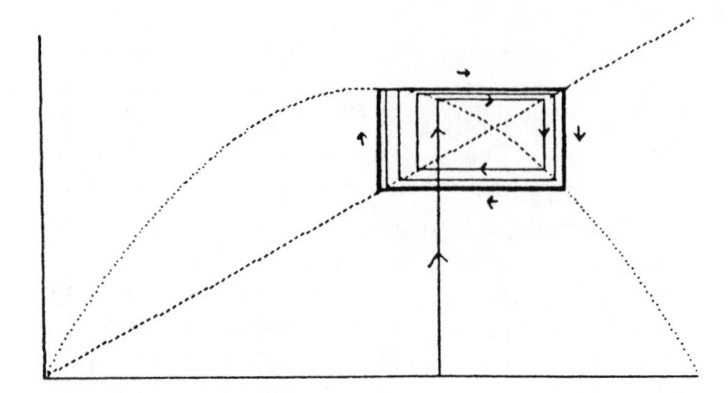

Figure 5. Cobweb graph for $X(n+1) = 3.2\, X(n)(1 - X(n))$.

For r larger than 3.45 this two–cycle becomes unstable, and a stable four–cycle emerges, another bifurcation. Figure 6 shows a stable four–cycle at $r = 3.5$ ($x_0 = .501$ is one of the points of the four–cycle). This pattern continues: As r gets larger, a stable 2^n–cycle emerges, and for a still larger r this cycle becomes unstable when a stable 2^{n+1}–cycle emerges. The set of points of a stable p–cycle is called a *periodic attractor*. Starting from a state close to this set, successive states will get closer to the set, every p units of time getting closer to the same member.

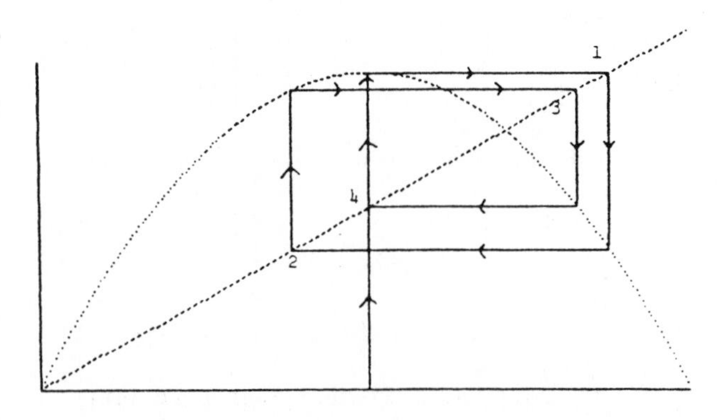

Figure 6. Cobweb graph for $X(n+1) = 3.5\, X(n)(1 - X(n))$.

Cobweb graphs reveal cycles and enable us to tell if they are attractors. In the case of a two–cycle we compute $X(n+2)$ in terms of $X(n)$. If $f(x) = rx(1 - x)$, then $X(n+1) = f(X(n))$ and $X(n+2) = f(f(X(n)))$. To graph $X(n+p)$ as a function of $X(n)$ and discover the p–cycles, f must be applied p times. In Figure 7 we have graphed the line $Y = X$ and the sixteenth–order polynomial corresponding to $X(n+4) = f(f(f(f(X(n)))))$ with $r = 3.5$. The four intersection points of downward slope, numbered 1 through 4, form a stable four–cycle. Two more intersection points between 2 and 4 and between 3 and 1 form an unstable two–cycle, and the remaining intersection point is an unstable equilibrium.

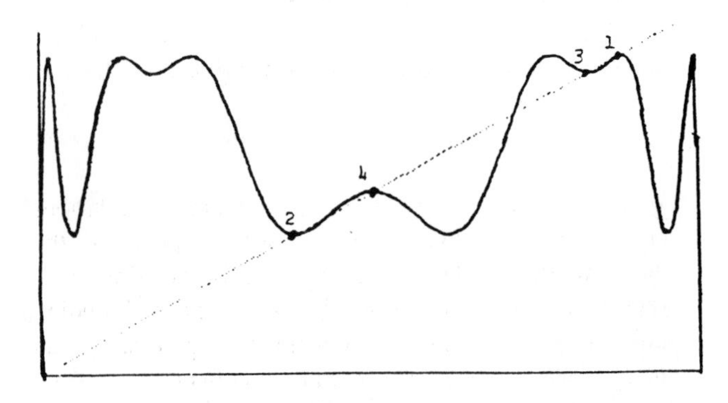

Figure 7. Points of a four–cycle.

A strange thing happens as r gets larger. The values of r at which successive 2^n–cycles emerge are all less than about 3.57. At approximately that value and beyond, infinitely many cycles are present. They are usually unstable, and the evolution of an individual state is quite complicated. CHAOS! In the last few years numerous examples of chaotic systems have been discovered primarily through computer investigation. Some theory has been built up [1], [3], [6]. Scientists find that these systems provide plausible models of such complex phenomena as the weather, certain chemical reactions, turbulence in fluid flow, dissipative biological systems, and irregularities in magnetic fields.

Discrete probability.

Probability theory for equally likely events depends on counting arguments which can be exhibited in tree diagrams. The basic ideas are the multiplication principle for multi–stage events and the addition principle for mutually exclusive cases. Counting of events can also be done by computer, e.g., by enumeration of branches in a tree.

These ideas fit in nicely with difference equations. For example, to count the number $A(n)$ of ways of drawing n marbles, with replacement and in order, from a bag containing k marbles, we first draw $n-1$ marbles in one of $A(n-1)$ ways. Then we draw one more marble in one of k ways. Thus, $A(n) = kA(n-1)$, and $A(n) = k^n$. To count the number $A(n,k)$ of ways of drawing exactly k red marbles in n draws from a bag containing 1 red and 1 blue marble, with replacement and in order, we consider two cases: Either the last marble is red (which can happen in $A(n-1,k-1)$ ways), or the last marble is blue (which can happen in $A(n-1,k)$ ways). Thus, we get the relationship of Pascal's triangle,

$$A(n,k) = A(n-1,k-1) + A(n-1,k),$$

the solution of which is ${}_nC_k$.

Applying probability theory and difference equations to genetics, we can examine such topics as the Hardy–Weinberg law, mutation, and selection. A nondeterministic computer simulation of a breeding experiment, based on a random number generator, can be compared with theoretical conclusions. With second–order difference equations we can also treat sex–linked genes and inbreeding.

Higher order linear difference equations.

An example of a higher order difference equation is the equation for the Fibonacci numbers: $F(n+2) = F(n+1) + F(n)$. The method for solving such equations is similar to that for differential equations: Construct the *characteristic equation* ($r^2 = r + 1$ in this case), find the roots, say r_1 and r_2, and write the general solution as $F(n) = a\,r_1^n + b\,r_2^n$, where a and b are constants depending on $F(0)$ and $F(1)$.

This method can be applied to the gambler's ruin problem. Suppose the probability of winning a dollar is p, the probability of losing a dollar is $q = 1-p$, and you play until you have 0 or N dollars. What is the probability $P(n)$ that you eventually go broke if you start with n dollars? There are two cases: Either you win the next bet and eventually go broke, with probability $pP(n+1)$, or you lose the next bet and eventually go broke, with probability $qP(n-1)$. Thus,

$$P(n) = pP(n+1) + qP(n-1), \text{ with } P(0) = 1 \text{ and } P(N) = 0.$$

The solution when $p \neq 0.5$ is

$$P(n) = (s^n - s^N)/(1 - s^N), \text{ where } s = q/p.$$

A simple computer program to compute $P(n)$ for different values of p, n and N teaches students valuable lessons about gambling, such as "If you mind losing, don't gamble".

If the roots of the characteristic equation are imaginary, then the general (real–valued) solution is a combination of powers of imaginary numbers. The graphs of these solutions exhibit oscillatory behavior, giving the students insight into imaginary numbers and, if you wish to mention it, their connection with trigonometric functions.

Other applications of higher order linear equations include a simple model of the economy and a model of a population with age structure. In both of these models roots of the characteristic equation can be real or complex. Solutions of the first for different values of the parameters reveal the conditions making the economy stable. The difference equation for the second model has order k, where k is the number of age groups in the population. In certain cases when the roots are complex, population waves occur, and each age group oscillates in size. This can be demonstrated by computer programs (see [7] and [5]).

Matrix algebra and applications.

After a standard introduction to vector and matrix algebra, we treat systems of difference equations,

$$X(n+1) = R\,X(n),$$

where R is a k by k matrix and $X(n)$ is a column vector. The solution is $X(n) = R^n X(0)$, and access to a package that does matrix multiplication is helpful for performing the iterations.

We may write solutions of the system in the form

$$X(n) = a_1 r_1^n V_1 + \cdots + a_k r_k^n V_k,$$

where $r_1, \ldots, r_k$ are the characteristic values, and $V_1, \ldots, V_k$ are the corresponding characteristic vectors. A nice application of this can be made to Markov processes, such as the random walk. For regular Markov processes, $r = 1$ is a characteristic value, and all others satisfy $|r| < 1$. Thus $X(n)$ converges to the (normalized) characteristic vector corresponding to $r = 1$. Computation shows that each column of R^n converges to this vector.

A second application is the matrix form of the age–dependent population model, in which the matrix R contains information on fertility and survival rates. When the characteristic value r of largest modulus is unique (and necessarily positive), it can be computed approximately as the limit of the sequence $x_1(n+1)/x_1(n)$, where $x_1(n)$ is the first component of the population vector $X(n)$. The rescaled vector $X(n)/r^n$ converges to a characteristic vector associated with r. Thus, while the population grows or decays exponentially, the ratios of population in different age groups become stable. In exceptional cases, when there are complex characteristic values of the same magnitude as r, oscillations occur [2].

A third application is a discrete version of the heat equation. Points on a bar or plate are given, and the temperature at each point is assumed equal to the average of the temperatures of neighboring points one time unit earlier. This yields a matrix system with a nonhomogeneous vector term contributed by boundary points, where the temperature is held constant. The characteristic values of the associated matrix R satisfy $|r| < 1$. The temperature profile converges to an equilibrium distribution, and successive profiles can be displayed on the computer.

Nonlinear systems of difference equations.

Examples of nonlinear systems include models of predator–prey relationships and interspecies competition for food. We linearize these systems near equilibria and study the stability of the resulting linear difference systems. When $X(n)$ and $Y(n)$ denote the numbers of predators and prey at time n, we obtain a cobweb–like graph by plotting the points $(X(n),Y(n))$.

Some nonlinear systems have no stable equilibria, for example, the system

$$X(n+1) = -1.4\, X(n)^2 + Y(n) + 1,$$

$$Y(n+1) = 0.3\, X(n).$$

Again we use the computer to help us understand this system. In Figure 8 we plot the points $(X(n),Y(n))$ for $0 \leq n \leq 2000$, starting from $X(0) = Y(0) = 0.1$. What results is the famous Hénon attractor, thought to be a nonperiodic, indecomposable attractor. Points initially close together get farther apart, so if we were approximating solutions, after a certain period of time our approximations would not be accurate. But we know that the solution is in this noodle–shaped region. There is good news and bad news. We can make general predictions (our state is in the region), but not specific ones (we have no idea where the state is in this region).

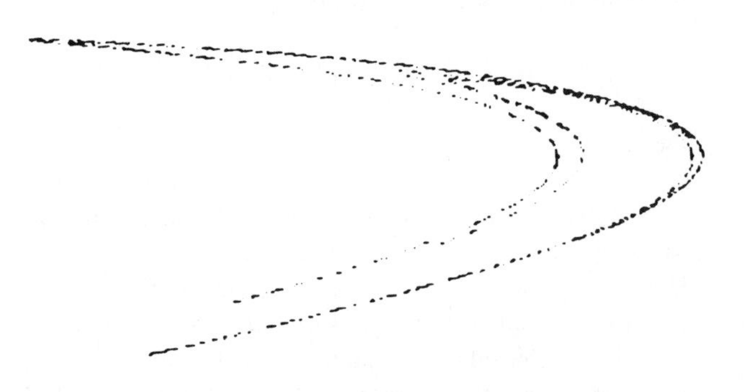

Figure 8. The Hénon attractor.

Could it be that the Universe as a whole satisfies a similar, but much more complicated, system of equations? By the Uncertainty Principle, the positions and momenta of particles can never be determined simultaneously with perfect accuracy. If the Universe has an attractor, we will never know where we are on the attractor. Being on or near the attractor means that there is general determinism, but not knowing where we are on the attractor implies that strict predestination is impossible. How many other mathematics courses can end on such a philosophical note?

Remark.

Copies of programs and other materials for teaching the course outlined here can be obtained from the Georgetown University Mathematics Department.

REFERENCES

1. Collet, P., and J.–P. Eckmann. **Iterated Maps on the Interval as Dynamical Systems**. Boston: Birkhauser, 1980.

2. Cull, P., and A. Vogt. "The periodic limit for the Leslie model." **Mathematical Biosciences 21** (1974), 39–54.

3. Devaney, R. L. **An Introduction to Chaotic Dynamical Systems**. Menlo Park, CA: Benjamin/Cummings, 1986.

4. Goldberg, S. **Introduction to Difference Equations**. New York: John Wiley & Sons, 1961.

5. Hoppensteadt, F. C. **Mathematical Methods of Population Biology**. Cambridge: Cambridge University Press, 1982.

6. Ruelle, D. "Strange Attractors." **Mathematical Intelligencer 2**, no. 3 (1980), 126–137.

7. Sandefur, J. T. **Discrete Mathematics with Finite Difference Equations**. Duplicated lecture notes, Mathematics Department, Georgetown University, 1983.

8. Ungar, P. "Review." **American Mathematical Monthly 93** (1986), 221–230.

The Use of Computing in the Teaching of Linear Algebra

Eugene Herman, Howard Anton, Alan Tucker, and Garry Helzer

[Editor's note: This is an edited transcript of a CCIME panel discussion presented at the meeting of the Mathematical Association of America in Laramie, Wyoming, in August, 1985. Eugene Herman was the organizer and moderator of the panel, and the other authors were panelists. Each author's contribution is identified separately.]

Herman:

When mathematics instructors began to have their students use calculators in precalculus and calculus courses, they were distracted by three phantom issues: the "fairness" issue of whether all students had equal access to the technology; the "legitimacy" issue of whether students ought to be allowed to use the devices during exams; and the "mindlessness" issue of whether students would lose valuable skills and become mere button pushers. These issues have led to more agonizing than useful insights; we are better off when we recognize them as phantoms that have largely dissipated over the years. We should focus instead on the following more substantial and productive questions:

1. What changes should we make in the topics we teach and how we teach them that will capitalize on the powerful calculating and computing devices available to our students?

2. Where can we find teaching materials, or how can we design our own teaching materials, that make good use of this technology?

For the topics in linear algebra, the outline of a response to these questions is taking shape. First, as all of us on the panel agree, linear algebra is an especially appropriate course in which to use the computer. There are several reasons for this. We teach linear algebra to many more students than we did thirty years ago, primarily because the computer has allowed scientists to make useful calculations based on linear models with large numbers of variables and equations. Therefore, we ought to show students some of these applications that have stimulated the enormous growth of interest in linear algebra. To do so we need the computer, as few interesting applications lead to small matrices that can be manipulated by hand. Another reason for using the computer is to enable students to carry out exploratory computations that allow them to discover patterns. Hand computations with matrices are so laborious and so likely to contain errors that students can rarely discover useful patterns with them. Furthermore, the hand computations are usually dull; their repetitive use teaches students little of value and instead persuades many that the subject itself is dull. Thus, to replace laborious, dull homework by more interesting and challenging assignments is yet another reason to use the computer.

Let us consider what these more interesting assignments might be and what the main topics in a sophomore linear algebra course might become. Surely the elementary theory concerning the concepts of linear independence, span, basis, and dimension will remain fundamental, and properties of the algebra of matrices and linear transformations will not lose their importance. The geometry of vectors will continue to provide important insights and examples. The analysis of linear systems of equations and the investigation of eigenvectors and eigenvalues, if changed at all, will require added emphasis, since the software enables one to ask so many interesting questions involving these objects.

However, reduction methods and echelon forms will need a different approach, since the software carries out the necessary computations by algorithms that are often quite different from those now taught. Perhaps reduction to upper triangular form, possibly using partial pivoting (and followed by back substitution if there is an equation to be solved), should be the main hand technique, as this is much like the *LU*–decomposition that a good software package uses. Perhaps there should also be some discussion of modern numerical linear algebra algorithms, although, since these are far from elementary, they ought not be the students' first taste of numerical methods.

Methods for computing matrix inverses and discussion of determinants probably require less attention. Applications rarely require the computation of a matrix inverse; it is almost always better to solve an appropriate linear system. Determinants decrease in importance because their main application — providing a formula for computing eigenvalues — will be less important; modern algorithms for approximating eigenvalues and eigenvectors make no use of determinants. Furthermore, two other applications of the determinant, Cramer's Rule and the adjoint formula for the inverse, become even more superfluous when one has powerful software to do the computations, software which makes no use of these formulas.

There should also be a subtle alteration in emphasis throughout the course. Rather than paying close attention to the entries of a matrix, a point of view that is reinforced by hand computation, we should emphasize properties of the matrix as an entity.

The major addition to the sophomore linear algebra course should be the study of applications. Some interesting applications that lead to linear systems of equations are: temperature distributions found by approximating values at discrete grid points, input–output models in economics, electrical circuit analysis, least–squares approximation, balancing chemical reactions, and network analysis. Applications that involve finding eigenvalues and eigenvectors include: Markov chains, Leslie population models, and models of genetic inheritance. If there is time to study first–order linear systems of differential equations, many more applications become accessible. These include flow of liquid between tanks, flow of substances between biological compartments, and systems of springs and masses.

To carry out such changes, both appropriate software and the right teaching materials are needed. By appropriate software I mean a package that enables students to carry out highly accurate matrix computations with a minimum of bother. These computational capabilities should include at least the following: solving a system of equations, finding the rank of a matrix, extracting a basis from a spanning set of vectors, finding all the eigenvalues and eigenvectors of a square matrix, and performing all the usual matrix algebra operations. A few such packages are already available, and more are likely to be available soon. Teaching materials, of course, will appear only later. However, prospective authors can resolve the second of my opening questions for the greatest number of instructors if they keep the following fundamental design principle in mind. Teaching materials should be designed to go with any appropriate software package, just as calculus and precalculus materials that require the use of a calculator are designed to go with any appropriate calculator.

[Editor's note: Gene Herman is the author of an "appropriate software" package, *MAtriX Algebra Calculator* (*MAX*), which runs on IBM PC's and compatibles. *MAX* is to be published soon by Brooks/Cole.]

Anton:

I would like to begin with some thoughts on a broader topic: computing and the teaching of mathematics. There is no doubt that we are in the midst of a genuine revolution in the teaching of mathematics resulting from the availability of powerful, inexpensive microcomputers. Moreover, college teachers are now experiencing the first generation of "computer babies", students who have grown up with computers, who regard their existence as a natural state of affairs, and who, in many cases, are highly skilled in their use before they enter college.

Motivated by this computer–hungry audience and seeking to justify expenditures on computer equipment, faculty have rushed to incorporate the computer into a variety of college courses as quickly as possible. Unfortunately, our understanding of the pedagogical role of this powerful new tool has lagged behind the hardware development, and much of the educational software being produced may be more detrimental than helpful in teaching mathematics. Why do I say this? It is my feeling that in mathematics the thinking and learning processes that take place in front of a computer screen are quite different from the traditional process in which the student reads a page of text and sits quietly with a blank sheet of paper thinking. In the former case there are visual stimuli constantly directing the thought process, whereas in the latter the mind is more free to wander, looking for an inspirational spark wherever it can be found. I believe that each of these learning methods plays a useful role which the other cannot duplicate. However, I also believe that, in the rush to bring the computer into mathematics courses, there has not been adequate attention paid to the pedagogical aspects of the software being developed and used. There has been a failure to recognize that a computer is not a textbook — it can do some things better and others not as well. On the other hand, the field of pedagogical software is in its infancy, and we should see this problem begin to resolve itself as "teachers" rather than "programmers" begin to exert a greater influence on the design of pedagogical software.

What should a teacher look for in pedagogical software before committing the students' valuable time to learning it and using it in a course? There are three obvious criteria. It must be:

1. WELL DESIGNED — it should be error free, user friendly, and easy to learn;

2. WELL WRITTEN — like a good textbook, it should explain or display ideas clearly;

3. PEDAGOGICALLY EFFICIENT — it should not duplicate a text; i.e., it should focus on topics and methods that are either beyond the limitations of the written medium, or, if the topic can be learned from a text, the software should teach it in <u>less time</u>.

I think the first two criteria are self–explanatory, so let me explain the third. Our students have a limited amount of time to devote to each course, and we as teachers must use that time appropriately so the students learn as much as possible. Learning to operate a software package, traveling to a computer, turning it on, and typing are essentially wastes of time in that they are only tangential to mastery of the mathematical subject matter. Unless that wasted time is offset by more rapid learning or a deeper understanding of the mathematics, the computer package is counter–productive and should not be used.

As a simple illustration, consider the problem of teaching the student that inscribed rectangles tend to fill out the area under a curve as the partition size diminishes. Most calculus texts discuss the idea and give three or four pictures to illustrate the point. It is my experience that even mediocre students can master this idea from such presentations. Nevertheless, I have seen a host of calculus software packages for the student that produce beautiful, dynamic pictures of this process. They are gorgeous to look at and make an excellent classroom demonstration, but it is hardly worth having the student learn to use the program, travel to the computer, turn it on, boot the disk, and do the keyboarding to learn a concept that the text can teach equally well in less time.

Let me turn now to the use of the computer in linear algebra. Of all the courses in the undergraduate mathematics curriculum, none cries out more for the computer. Anyone who has taught sophomore–level linear algebra knows that it is extremely difficult to cover all the important material in a single semester, let alone touch on applications. Part of the problem is that an inordinate amount of the student's time is spent solving linear systems and reducing matrices to echelon form well after that serves any useful pedagogical purpose. For example, if the student knows how to set up a homogeneous linear system to determine whether a set of vectors is linearly independent and knows that the existence of nontrivial solutions implies linear dependence, then what is the purpose of requiring the student to solve the system? It would certainly be more efficient to have a computer do the computations, leaving the student free to focus on the method of solution. The potential savings of time are enormous. It is not unreasonable that a well designed computer package might save as much as 50% of the student's homework time, making it possible to assign many more exercises and move more quickly through the course — a prime example of pedagogical efficiency.

In 1984 I began designing a microcomputer package called LINEAR–KIT for use by students as a problem–solving tool in undergraduate linear algebra courses. My objectives were fourfold:

1. I wanted a pedagogically efficient package that would enable the student to solve the standard homework problems in less time, thereby allowing the instructor to cover more material or the same material in more depth.

2. I wanted it to be usable with virtually no instruction.

3. I did not want a purely decimal arithmetic package. I wanted the capability to do rational arithmetic exactly to preserve theoretical relationships. I also wanted to be able to solve linear systems with infinitely many solutions, producing the appropriate parameters in the answer. My objective was a package that could be used to explore theorems as well as obtain numerical answers.

4. I wanted the package to fit easily into existing linear algebra courses and require relatively little keyboarding of data by the student. To achieve this I have developed a booklet of linear algebra problems that can be solved with LINEAR–KIT. The booklet is accompanied by a data disk that contains all the matrices and vectors needed for the problems.

I am not alone in the development of such packages. Eugene Herman's MAX package, used successfully at Grinnell College, and the extremely powerful MATLAB software produced by The Math Works are two other packages that are already available. The software is here and the time has come to begin restructuring our undergraduate linear algebra courses around these new tools.

[Editor's note: The *Linear–Kit* software, *The Linear–Kit Problem Book*, and *The Linear–Kit Applications Problem Book* are available from John Wiley and Sons.]

Tucker:

I have spent the last two years developing a new course at SUNY–Stony Brook that attempts to give students a unified introduction to the models, numerical methods, and theory of modern linear algebra. Linear models are now used at least as widely as calculus-based models. The world today is commonly thought to consist of large, complex systems with many input and output variables. Linear models are the primary tool for analyzing these systems. This course is gaining a reputation among many math students at my school of being the most useful mathematics course they take. With this goal of usefulness in mind, the material is presented with an eye towards making it easy to remember, not just for the next test, but for a lifetime of diverse uses. Underlying all aspects of this course is the role of computing.

Linear algebra is an ideal subject for a lower–level college course in mathematics because the theory, numerical techniques and applications are interwoven so beautifully. The theory of linear algebra is powerful, yet easily accessible and even likable. It simplifies and clarifies the workings of linear models and related computations. This is what mathematics is really about: making things simple and clear. The theory provides important answers that go beyond results we could obtain by brute computation. But without the computations, the theory would be unmotivated and dry. For too many students mathematics is either a collection of techniques, as in calculus, or a collection of formal theory with limited applications, as in most courses after calculus (including traditional linear algebra courses). My course tries to remove this artificial dichotomy.

The field of numerical linear algebra is very young, having been dependent on digital computers for its development. Yet it has wrought major changes in what theory should be taught in an introductory course. The standout example of a modern linear algebra text is G. Strang's *Linear Algebra and Its Applications.* Once the theory was needed as an alternative to numerical computation, which was hopelessly difficult. Now theory helps direct and interpret the numerical computation, which computers do for us. Today's close links between linear algebra and computers are nothing new. Around 1850 in England, when Cayley first developed matrix multiplication and the basic theory of linear algebra, Babbage was at the same time designing the first modern mechanical computer. More recently, Alan Turing, who was the father of computing theory, was also the person who discovered the *LU* decomposition, a central result of numerical linear algebra.

The applications of linear algebra are powerful, diverse and easily understood — with the help of some numerical experimentation on computers. I introduce students to economic input–output models, population growth models, Markov chains, linear programming, computer graphics, regression and other statistical techniques, numerical methods for approximate solutions to calculus problems, and more. These diverse applications reinforce each other and the associated theory. Indeed, without these motivating applications, several of the more theoretical topics could not be covered in an introductory text.

My course develops linear algebra around matrices. Vector spaces in the abstract are barely considered. The course puts problem solving and an intuitive treatment of theory (based on numerical experience) first, with a proof–oriented approach reserved for a second course, the same way calculus is taught. The text for the course has a straightforward organization: Chapter I has introductory linear models, Chapter II has matrix algebra, Chapter III develops different ways to solve a system of linear equations, Chapter IV develops applications, and Chapter V has vector space theory associated with matrices and topics such as pseudo–inverses and orthogonalization.

Many linear algebra courses start with Gaussian elimination before any matrix algebra. I first pose problems in Chapter I, then develop a mathematical language for stating and recasting the problems in Chapter II, then look at ways to solve the problems in Chapter III. Four different solution methods are presented with an analysis of strengths and weaknesses of each.

In applications of linear algebra, especially those involving heavy computation, the most difficult aspect is usually understanding matrix expressions, such as $Ue^{D}U^{-1}$. Students in a traditional course have little preparation for understanding such expressions. My course constantly forces students to interpret the meanings of matrix expressions at the same time they perform matrix computations. Matrix notation is used as much as possible, rather than frequently writing out systems of equations. The material is unified pedagogically by repeated use of a few linear models to illustrate all new concepts and numerical techniques. These models give the student mental pictures to "visualize" new ideas and computational methods, and they help the student remember concepts and techniques after the course is over.

Although this course is often informal ("proving theorems" by numerical example), I am able to cover topics normally left to a more advanced course, such as matrix norms, matrix decompositions and approximation by orthogonal polynomials. These advanced topics are easy to motivate numerically, and they find immediate, concrete applications. In addition, they are finite–dimensional versions of important theory in functional analysis; for example, the eigenvalue decomposition of a matrix is a special case of the spectral representation of linear operators.

Today there is a major curriculum debate in the mathematics community between computer science-oriented discrete mathematics and classical calculus-based mathematics. Linear algebra, especially as viewed in this course, is right in the middle of this debate. (Linear algebra and matrices have always been in the middle of such debates: Matrices were a core topic in the best known first–year college mathematics text before 1950, Hall and Knight's *College Algebra*, and much of Kemeny, Snell and Thompson's *Introduction to Finite Mathematics* involved new applications of linear algebra: Markov chains and linear programming). My course attempts to present a healthy interplay between mathematics and computer science, between continuous and discrete modes of thinking. The complementary roles of continuous and discrete thinking are typified by the different uses of the euclidean norm and sum norm. An important example of computer science thinking is the role of matrix representations such as the *LU* decomposition. They are viewed as a way to preprocess the data in a matrix to solve quickly certain types of matrix problems. Computer science also gives insights into the teaching of linear algebra: A computer scientist's distinction between high–level and low–level languages applies to linear algebra proofs; a high level proof involves matrix notation, while a low–level proof involves individual entries.

[Editor's note: The text *A Unified Introduction to Linear Algebra: Models, Methods and Theory* that Alan Tucker wrote for this course will be published by MacMillan in summer 1987.]

Helzer:

The first question that comes up in a discussion of introduction of computers into the linear algebra course is: Will the new material help students learn linear algebra? But what is linear algebra? The linear algebra course the questioner has in mind is usually the axiomatic, coordinate–free approach to linear algebra that has been popular for the last quarter–century. The computer, at least as I use it, cannot help students learn axiomatics. It can provide examples to guide intuition and give the student a feeling for what might or might not be true, but it cannot teach the processes of axiomatic reasoning and rigorous proof construction. What the computer can do quite effectively is provide a calculator for nontrivial examples.

In the last twenty years matrices and matrix computations have become increasingly important in many fields; today the audience for a linear algebra course includes much more than mathematics majors or even science majors. This change is due to the universal use of computers. Whatever the origin of the particular problem, when numbers are put into a computer, some matrix algebra is usually involved in the computation. There is a need for an introductory linear algebra course to serve this new audience the computer has created.

This audience is not well served by the traditional linear algebra course. The main problem, as I see it, is that the linear algebra course is likely to be the last formal mathematics course these people take, and what they learn will be applied in a computational context. The course should be designed with this in mind. For example, machine calculations are inexact, but theoretical discussions of machine arithmetic are difficult and out of place in an introductory course. Instead, computer exercises can be designed to emphasize the inexact nature of machine arithmetic and emphasize that judgment and a certain measure of common sense are necessary when interpreting machine computations.

Here is an example of what I mean. Suppose that in performing a row reduction on a computer, the first row is used to set the entries in the first column and the subsequent rows to zero. In production code, these zeros are not usually stored and auxiliary information is stored in their place. In my course, on the other hand, the routines used to row–reduce matrices also carry out the zero producing computations and store the resulting zeros in their proper place. When the results are printed out, these "zeros" are often non–zero! Roundoff error has crept in. The student must then decide if the non–zero entries are just zero plus a little roundoff error or if something has gone wrong with the computation.

In my course the computer is not used to help students learn axiomatics, nor is it used to teach them to write serious computational programs. It is used to:

1. provide a tool for solving linear algebra problems, a tool the student can use after leaving the course (and some do so use it);

2. give some experience, but not a theoretical grounding, in the inexact nature of machine computations.

When I first began working on the course, I wrote down three problems I wanted the students to be able to solve numerically, at least in well–conditioned cases. These were: solving linear systems, computing eigenvalues, and solving least–squares problems, all of "modest" size. These goals are reached by most of the students as far as the linear systems and least squares problems are concerned. For the eigenvalue problem, I cover numerical methods only for the symmetric case. (The treatment of the non–symmetric case using the characteristic polynomial is also covered, but this approach is good only for small or carefully chosen examples.)

The computer assignments I give differ from ordinary homework problems only in that the central computation (row reduction, orthogonalization, eigenvalue computation) is done on the machine. The machine output is never the final answer; it must be interpreted and explained. Thus, the computer is used for linear algebra calculations in the same way that a calculator is used in a physics or engineering course. I keep the assignments short to minimize keyboard time. In the course of a semester, I assign and collect ten to twelve computer assignments, and I weight them heavily enough to insure that the students will do them.

For example, in an early assignment the student calls up (or types in) a matrix A and requests its row–reduced echelon form by typing "echelon A." From this, the student must then answer the following questions. What is the dimension of the row space of A? The column space of A? The null space of A? What set of columns forms a basis for the column space of A? If A is the augmented matrix of a linear system, what are the solutions?

To function as an effective linear algebra calculator for such problems, the software needs to have the following features:

1. Matrices should be independent of programs.

2. Standard matrix arithmetic should be quick and easy.

3. High–level programming (macros) should be available.

By 1. I mean that the user should be able to sit at the keyboard and quickly type in and save several matrices or, alternatively, call up matrices already stored in the system, perhaps by the instructor. This is in contrast to, say, BASIC or Pascal, in which a program must be written to dimension a matrix and then store the appropriate numbers in the rows and columns.

Quick and easy matrix arithmetic means that a user should be able to add, subtract, transpose, multiply, invert, and store the results for later retrieval, each with a few keystrokes. Most symbolic manipulation programs, such as muMath, are not sufficiently fast for this.

By 3. I mean that the software should permit the user to define new operations as sequences of operations already defined. This requirement, however, is optional in an introductory course, as it can be a time–consuming feature for students to learn to use. But even then it can be useful to the instructor.

The best choice of software to meet these requirements probably depends on local circumstances, including the personnel and hardware available. I advocate using a general–purpose computer language, APL. Although the effort invested in learning the system is greater than with special–purpose software, the payoff is much greater, because the same techniques can be applied outside the linear algebra course. In fact, there is a growing body of APL software available for mathematics. Furthermore, APL is a language that is especially well suited for use as a linear algebra calculator. It provides a workspace environment for handling collections of matrices, and its primitive data structures are vectors, matrices, and higher dimensional arrays. Many of the matrix operations are built in.

The introduction of computers in the linear algebra course changes the syllabus in the following ways:

1. Matrices and matrix algebra acquire more importance than is usually the case in an axiomatic course.

2. Some remarks about roundoff error and pivoting strategies need to be added. I confine this material to a few rules of thumb and design the routines to show rather than suppress the roundoff error that occurs.

3. Several standard algorithms do not work well on a computer. The easiest way for a student to compute a determinant is not usually row reduction, because of the fractions involved, but it is the fastest way on a computer. Programming expansion by minors leads to long running times for matrices of only modest size. It takes some time to explain these facts.

4. The standard Gram–Schmidt process does not work at all well on the computer. Very inaccurate results can be obtained by applying it to three vectors in three dimensional space. It is best to replace it by the Householder algorithm, especially in a service course, as Gram–Schmidt is almost never used in practical situations. It takes more time to explain the Householder algorithm and the associated QR factorization.

5. Finding eigenvalues as roots of the characteristic polynomial does not work well on the computer. The numerical routines used to compute eigenvalues do not involve the characteristic polynomial.

6. In a service course one should include some discussion of the least–squares problem.

7. Items 2 through 6 lead to inclusion of material not found in the traditional linear algebra course. Clearly something must go. The usual result is de–emphasis of abstract vector spaces and axiomatics, confining most of the material to subspaces of R^n.

[Editor's note: Garry Helzer has written a textbook for his course at the University of Maryland: *Applied Linear Algebra with APL*, Little Brown, 1983.]

Using Computers in Teaching Linear Algebra

Morris Orzech

The goal of this article is to give an overview of approaches to the use of computers in teaching linear and matrix algebra. Most of these approaches would make sense in the context of other subjects as well, but in discussing the tools and their applications I will limit myself to linear algebra.

I begin by describing the main computer–related tools I have used or whose use by others I am aware of through personal contact or reading. The use of a particular tool may well be a reflection of an underlying heuristic philosophy. Hence discussion of how the tools are exploited is largely omitted in the initial catalog and postponed to the next section, which deals with ideas behind computer use. Finally, I offer some comments on how the suitability of a given approach may be influenced by the instructional setting.

Computer tools for linear algebra.

As a matter of convenience, I group the tools being described in three categories: 1. calculating tools; 2. algebraic manipulators; 3. computer languages. These groups overlap, and there may be room for argument as to which category a given tool represents.

1. Calculating tools. My experience over the last five years, and that of others using the computer for linear algebra instruction, involves a variety of programs that perform matrix operations, thus relieving the user of the tedium of hand–calculations. At Queen's University, students have access to two very simple and two more sophisticated programs. The simplest of these is Ematrix, written in BASIC (and subject to student modification); it asks for matrix input and row–reduces the matrix. A program called Qmatrix (because it handles rational numbers as input and output) provides the student with a template for entering and editing a matrix. It does only row–reduction and inversion. A more comprehensive program, Linalg, runs under APL and provides a more extensive set of operations on a single matrix. It is menu–driven, provides the conveniences of rational number output and of a template for matrix input, and can be modified by the student who knows APL.

The three programs mentioned so far are available to students at terminals to the mainframe academic computer. A program named Matrixpad is available to students who have an IBM–compatible personal computer. (Most incoming engineering students now purchase a Zenith microcomputer, and clusters of computers are available on campus.) Matrixpad also facilitates entry via a template, and it provides rational or decimal operating modes. It is driven by single–key commands (like a calculator) and can do both unary and binary operations. It is limited in size of matrices and in its ability to extract eigenvalues. Its features cannot be modified by the student, but it is robust and has built–in help screens and informative error messages.

I have knowledge of two packages prepared at other universities for student use in linear algebra courses. At Cornell, S. U. Chase and R. Connelly, Jr., developed an APL–based program, RMR, which can do elementary row operations under user guidance and (optionally) provide information about the row and column spaces associated with a matrix. The most comprehensive program I am aware of is Matrix, designed by J. J. Cannon at the University of Sydney. Its use at the universities of Sydney and Melbourne was described by Newman [8]. The program provides a matrix calculator with rational or real entries, a command language, access to prepared problems, and a facility for teachers to write exercises, hints, and solutions.

I will also mention PC–MATLAB and Scientific Desk, two tool kits which can handle matrices and do much more. Reviews of these may be found in the July and September 1985 issues of *SIAM News*, but I am not aware of their being used for linear algebra instruction.

2. Algebraic manipulators. The algebraic manipulation systems I have seen mentioned in relation with university teaching are MACSYMA, Maple, muMath, and Reduce (see [11] and the article by Hosack in this volume). Although linear algebra has not been the sole focus for the use of these versatile programs, it is one of the areas for which applications have been made or suggested. These programs permit work with matrices whose entries are rational numbers or algebraic expressions. They differ in what functions are built in, but all allow extension through user programs in a language specific to the system. MuMath is available for a variety of microcomputers; the other systems are mainframe or minicomputer based.

3. Computer languages. In addition to the languages provided with symbolic algebra systems, APL is a language naturally linked to mathematics and particularly to matrix algebra. It provides an interactive calculator with extensive and versatile functions for array manipulation and user–defined functions as accessible as primitive ones.

If a teacher's philosophical bent dictates the use of a computer language as a tool in teaching linear algebra, then BASIC and Pascal are natural contenders because of their widespread use, and in the case of Pascal, because its syntax is sometimes relevant to mathematical needs and conventions [10].

Philosophies behind computer use.

In describing philosophies underlying use of computers, one can focus on the computer's role; for example, is it to teach by instructing the student or by being instructed? On the other hand, one can relate use of the computer to a model of how students learn, analyzing the computer's potential for developing and integrating conceptual and computational foundations. The latter analysis is one on which I have seen little discourse, particularly vis–à–vis college mathematics. I will therefore restrict myself to describing possible educational roles for the computer, only incidentally touching on psychological foundations for choosing a role to exploit.

I will touch upon five overlapping ways in which people have used computers as an aid in teaching linear algebra: 1. as a provider of tutorial assistance; 2. as a tool for implementing illustrations and practical applications; 3. as a device for experimentation and discovery; 4. as a means for creating a self–paced learning environment; 5. as a context for describing and illustrating mathematical concepts.

1. Computers for tutorial use. In the previous section I referred to the program Matrix, which allows the student to access prepared examples and the instructor to prepare solutions and hints. The RMR program may be thought of as a tutorial in the limited area of Gaussian elimination. Matrixpad provides a similar capability as one of its functions, along with a facility for teaching students to distinguish elementary row operations from non–elementary ones they often attempt.

My experience, and that at Cornell with RMR, is that this sort of tool is effective in its narrow area of application. Students use it to learn the mechanical technique, and they are less apt to need the teacher's help with the technique or to err in applying it.

2. Computers for illustration. My own experience in this area is limited to using Matrixpad to generate more natural and sophisticated examples than was my wont in the past. Beyond this my comments are reflections of conversations with colleagues, my own musings, and extrapolations from recent writings about computers in education.

An area in which computers can provide hitherto unattainable flexibility is graphical displays. Given that algebraic and geometric outlooks play a synergistic role in linear algebra, the idea of exploiting computer graphics in this subject is most attractive. In the words of one commentator [5]

> "[The availability of computers] will also make it possible to motivate many topics by examples of immediate significance to the student. For example, elementary linear algebra can be motivated and demonstrated by the need to perform coordinate transformations, rotations, and scaling in the graphical display of objects."

Although the technical capability for the realization of this vision may be available, it has not to my knowledge been realized in a program that is accessible and affordable for class use.

3. Computers for experimentation and discovery. Symbolic algebra programs have been seen by some people as having a special role to play in this area. Lane [7] discusses this, offering examples of how MACSYMA or muMath might be used:

> "Use your computer algebra system to form the matrix products
>
> $$\begin{bmatrix} a & 1 \\ 1 & 0 \end{bmatrix}\begin{bmatrix} b & 1 \\ 1 & 0 \end{bmatrix}, \quad \begin{bmatrix} a & 1 \\ 1 & 0 \end{bmatrix}\begin{bmatrix} b & 1 \\ 1 & 0 \end{bmatrix}\begin{bmatrix} c & 1 \\ 1 & 0 \end{bmatrix}, \text{etc.},$$
>
> each time including one more matrix until you can infer the general form of the elements in the product. Then see if you can use the system to help inductively prove your general form."

Computer algebra systems all provide symbolic manipulation capabilities more than adequate for linear algebra. But even the numeric tools described earlier can be used for experimentation. No matter which manipulative tool is adopted, successful exploitation of the calculator requires direction from the instructor and willingness on the part of the student to spend time on the process. In a linear algebra course I taught, where the students were merely introduced to Reduce, to APL, and to the Ematrix and Qmatrix calculators, the latter two programs got considerable use, APL less, and Reduce even less.

4. Computers for self–paced learning. When the notion of using computers in the learning environment first surfaced, a good deal of discussion focused on computer–based instruction. The idea now seems quiescent, but it manifests itself implicitly. In my experience students given computer–based materials use them at different (even unexpected) times and with varying intensity.

The hardware to make self–paced learning feasible via the computer is on the horizon in the form of optical disks which can hold integrated text, graphical material, and interactive programs. When the hardware becomes available we will have a better setting to test the promise of computer–based teaching and learning, but I am not aware of any existing linear algebra courses which provide a large component of self–paced computer–based instruction.

5. Computers for a context. This is an area in which there is a good deal to report, and which touches on issues of mathematics education over which there has been recent ferment. Even when all one has in mind is a simple layering of computer use over an existing course, content and presentation are influenced. More fundamental effects are likely where the goal is integration of computing skills and computing science knowledge into mathematics courses. I will take up three strategies for involving the use of computers: a. using available tools in an integral way; b. structuring a course around a particular tool, specifically APL; and c. viewing the linear algebra course as one part of a program which elsewhere exposes the student to computing science concepts and tools.

a. The engineering faculty at Queen's has initiated a program along the lines of those at Stevens Institute of Technology, Clarkson College, and others, wherein each student is assumed to have a particular personal computer for course use. The linear algebra program made available to our students was Matrixpad. My experience was that having this tool, *and being able to assume the students had access to it*, had a substantial influence on what I presented and how long I spent on it. Not only was I able to present more examples and more interesting ones, but I could also suggest to the students how they might generate their own.

More crucially, I found that demonstrating how the calculator's functions could be exploited to do various problems was an algorithmic way of reviewing material presented conceptually in the text. These complementary approaches were not possible when a good deal of time had to be spent on the mechanics of arithmetic and students could not reproduce my work.

b. Many mathematicians have a special fondness for APL and view it as a convenient tool for handling arrays of numbers. Some users of APL see it as a language which can be the foundation for describing mathematical operations arising in various contexts, particularly in linear algebra. At a cost of having to forego traditional notation (and perhaps some related conceptual notions, such as viewing matrix product as multiplication in a ring) one can thus hope to integrate the concepts in a course with algorithms for implementing them. Two approaches to this idea may be found in books by Helzer [6] and Sims [9].

c. In many situations where linear algebra is taught, most if not all students have some computer language at their disposal and have been or are being exposed to computing science concepts. One of the challenges that mathematics and computing science teachers face is to exploit the student's knowledge and the links between the disciplines.

The student's ability to write programs can be used for various purposes, e.g., to build a toolkit or to test understanding of a concept by implementing it. (Inarticulate writings can sometimes mask real understanding, sometimes suggest an understanding that isn't really there.) Dubinsky [3], [4] has proposed using computer processes and architectures as a way to describe mathematical concepts — composition using pipelines, for example. My experience suggests that these approaches benefit some students, but they are difficult to implement in a heterogeneous group where adherence to a given syllabus is a requirement. It is interesting to note the perspective of a high school educator [2]:

> "Good programming habits induce good intellectual habits (especially mathematical ones), and students who have learned to work within the confines of a formalized

programming language have an easier time with some of the more abstract aspects of mathematics."

The instructional context.

The use of computers in teaching linear algebra is not a new phenomenon, but there is a current wave of interest sparked by the advent of affordable personal computers. This does not mean that personal computers are the best means for incorporating computing in a course, but it does point to the fact that the context for use of computing tools, more than the tools alone, determines what will be instigated, done, or accepted by teachers and students to exploit computing power.

Contrasts occur on various fronts: The distribution of computing facilities may be via terminals, clusters of personal computers, or individually owned ones; a linear algebra course may be offered to students with or without background in computing science; the student body may be homogeneous or varied, accepting of their need for the subject or not; the faculty may be eager or reluctant to modify the curriculum with computer use in mind; and finally, student, faculty, and institutional financial resources may be ample or strained.

For each of the contrasts offered there may or may not be a choice that an instructor or a department can exercise; if there is, differing opinions are likely to emerge. As an example, we quote two views (admittedly not in context) as to what hardware is suitable and why [1], [12]:

> "Recognizing the cardinal importance of the individual and of the active involvement of each and every student in the computational activity ... we advocate the use of mathematical laboratories, based upon a set of microcomputers."

> "... we do not regard microcomputers as a very useful hardware base for the involvement of computers in undergraduate mathematics, given their current level of computing power."

Curriculum diversity also complicates the choice of strategy. While it may have been natural to decide at Stevens Tech that students should own personal computers, and that a course called Matrix Algebra with Computers should be offered, such a decision might not be appropriate elsewhere.

It is impossible from my limited perspective to make comfortable value judgments on particular approaches and tools. I see the diversity which now exists and the thrust towards integration of computing and mathematics as positive things. My own experience and that of others (see [8], for example) is that any given approach is unlikely to capture the interest of all students or all teachers. But with the variety of tools and approaches that seem to be under development and investigation, it seems safe to predict that, ten years from now, involvement of computers in subjects such as linear algebra will be a fact for everyone, even if the form of the involvement varies.

REFERENCES

1. Breuer, S., and G. Zwas. "The Computational Lab is a Valuable Tool." **SIAM News**, Sept. 1984, p. 8.

2. Cuoco, A. A. "On Merit Pay and the Use of Computers in Education." **SIAM News**, July 1984, p. 6.

3. Dubinsky, E. "Computer Experience as an Aid in Learning Mathematics Concepts." In **The Influence of Computers and Informatics on Mathematics and its Teaching**, Supporting Papers, Institut de Recherches sur l'Enseignement des Mathematiques, Strasbourg, 1985, pp. 61–70.

4. Dubinsky, E. "On the Links Between Teaching Mathematics and Computing." **SIAM News**, July 1985, p. 6.

5. Gear, C. W. "Let's Not Let Computers Upset the Apple Cart." **SIAM News**, March 1984, p. 6.

6. Helzer, G. **Applied Linear Algebra with APL.** Boston: Little, Brown & Co., 1982.

7. Lane, K. D. "Symbolic Manipulators and the Teaching of College Mathematics: Some Possible Consequences and Opportunities." In **The Influence of Computers and Informatics on Mathematics and its Teaching**, ibid., pp. 179–184.

8. Newman, M. F. "Some Software for Teaching Algebra." In **The Influence of Computers and Informatics on Mathematics and its Teaching**, ibid., pp. 195–197.

9. Sims, C. C. **Algebra, A Computational Approach**. New York: John Wiley & Sons, 1984.

10. Rall, L. B. "Pascal for Scientific Computation." **SIAM News**, March 1985, p. 8.

11. Squire, W. "muMath System Effective Tool for Algebra." **SIAM News**, November 1984, p. 4.

12. Thorne, M. "Computers as a University Teaching Aid: Towards a Strategy." In **The Influence of Computers and Informatics on Mathematics and its Teaching**, ibid., pp. 49–55.

Differential Equations Software Reviews

Howard Lewis Penn

[Editor's Note: Howard Penn was the organizer of a CCIME panel discussion on the use of computers in the teaching of differential equations; the panel was presented at the January, 1987, annual meeting of the MAA in San Antonio. The other panelists were J. M. A. Danby of North Carolina State University and Donald Lewis of Cornell University. Penn's presentation was based on an extensive survey of differential equations software, some of which was demonstrated, some described verbally. This article contains "capsule" reviews of a selected portion of that software; an extended version will appear elsewhere. Danby's article in this volume is an extended version of his panel presentation. We regret that Lewis' presentation on using a computer algebra system in teaching differential equations could not not be included.]

I review here some of the better differential equations software packages. Space precludes the inclusion of all available programs; see [2] and [3] for listings of other software available for differential equations. These reviews are based on limited use of each package and should be considered first impressions only.

Phaser by Hüseyin Koçak, Brown University. This package is available for MS–DOS computers (IBM PC's and compatibles) with CGA card or the equivalent; it will not run with a Hercules card or an EGA card. The program comes with a 224 page book titled *Differential and Difference Equations through Computer Experiments.* Packaged with the book are two disks, one for machines with an 8087 math coprocessor, the other for machines without. (Having the coprocessor makes a big difference in speed — it's highly recommended.) This package is far too extensive to describe in the space available; see [1] for a more complete review of both book and software.

Phaser is completely menu driven. It has over 60 differential and difference equations built in, and each of these has parameters that may be set. Therefore, just about any equation one might wish to look at is already available, but the user can also define equations. The output is graphical and very flexible. Among the plots available are direction fields, trajectories, and phase planes. Three dimensional plots are available for simultaneous equations, and these graphs can be changed by rotation about any axis. The graph in front of a given plane can be shown in one color, while that behind it is shown in another.

Phaser is a very powerful package and a real bargain. It is available for $44 from Springer–Verlag, 175 Fifth Ave., New York, NY 10010.

MacMath by John H. Hubbard and Beverley H. West, Cornell University. This package has versions available for the Macintosh and for the IBM AT with math coprocessor and EGA card. It is menu driven and allows the use of Euler, improved Euler, or fourth order Runge–Kutta methods to solve first order equations for one, two or (autonomous) three–dimensional systems.

The solutions are presented in well designed plots, and periodic nonautonomous systems receive especially nice treatment. The "Planets" routine solves the n–body problem for $2 \leq n \leq 10$, quickly enough to be fascinating; zero mass planets are allowed. Other routines find eigenvalues and eigenvectors of matrices by the Jacobi method or *QR* algorithm (512K Mac only), locate and identify bifurcation points for autonomous planar systems, and compute partial sums of Fourier series.

MacMath is a powerful package that would interest anyone involved in geometric study of differential equations. A pre–release version of the MS–DOS version did not run on a Zenith AT–clone with math coprocessor and Tecmar EGA card; it may be necessary to check out the package on the particular configuration on which it will be used. It is currently being sold by the authors for $40; write to MacMath, Department of Mathematics, White Hall, Cornell University Ithica, NY 14853. An agreement is being negotiated with Springer–Verlag, and there is a tentative price of $80 with textbook included.

MATT Graphical Display Package and Differential Equation Solver by Ben Staat, Scott F. Porter, and Mike Stark, Harvey Mudd College. MATT is a powerful differential equations solver and graphics package that has been adapted for IBM PC's and compatibles from a much larger software package designed for VAX computers. As of June, 1987, it was in its "Pre–Release Version 0.99," very close to release.

It can be ordered with or without math coprocessor support (8087 or 80287, depending on the computer); as with other computationally intensive packages, the coprocessor is highly recommended. It comes with device drivers for any of the IBM graphics modes (CGA 2– or 4–color or high res; EGA standard, monochrome, or medium res; PGA), but not for Hercules or other non–IBM protocols. There are also device drivers for several printers and plotters, but the standard versions of GRAPHICS.COM will dump screen images to a printer.

The user interface is similar to that of a symbolic manipulator (computer algebra system), in that problems are solved by entering instructions. The program can solve first order differential equations and systems of first order differential equations. It can be used to solve higher order equations by converting them to systems of first order equations. The syntax is easy to use, and the input is checked for errors. Solutions are computed by very accurate methods developed at Lawrence Livermore Laboratories. The results can be saved in a file for later use; therefore, the package might be used to produce demonstrations to be shown in class.

MATT is available for $50 from the Mathematics Department, Harvey Mudd College, Claremont, CA 91711. Eventually it will be marketed through some vendor.

Ordinary Differential Equations by J. L. Van Iwaarden, Hope College. This package is designed for use with the author's textbook, *Ordinary Differential Equations with Numerical Techniques* (Harcourt Brace Jovanovich, 1985); the book is not included with the software and is not required for effective use. An Apple II version is already available, and an MS–DOS version is soon to be released. There is a version available for DEC equipment from DECUS (free to members of DECUS). A more detailed review of the Apple version appears in [4].

The package is a menu driven set of programs that implement 17 numerical methods to solve differential equations. These include Euler, improved Euler, and Runge–Kutta for first and second order equations, Milne and Hamming predictor–corrector methods, Bessel methods, 3 series methods, and 3 methods for systems of equations. When a method is selected, the user is asked to enter a differential equation. The program graphs the direction field (where meaningful) and/or the solution of an initial value problem. It can overlay graphs of solutions to the same differential equation with different initial conditions, but it cannot overlay graphs generated with either a different number of terms or a different method. This package is useful for illustrating the different numeric techniques.

The Apple and MS–DOS versions are available for $75 from CONDUIT, The University of Iowa, Oakdale Campus, Iowa City, IA 52242.

Differential Equations Graphics Package by Sheldon Gordon, Suffolk County Community College. This package has versions for MS–DOS computers with graphics adapter and for TRS–80. It includes programs for linear first and second order differential equations; Euler approximation; graphical comparisons of Euler, improved Euler, and Runge–Kutta methods; tangent fields; logistic differential equations; Frobenius series; and Fourier series. A unit still in development will include graphical versions of some systems of equations, such as the Lotka–Volterra predator–prey system.

The linear first and second order programs allow construction of a non–zero right–hand–side from a menu of components; all the other programs have this feature and also allow the user to enter the desired expression directly. The Euler approximation program allows equations of the form $y' = f(y)$ and plots several approximations for different numbers of intervals. The tangent field program plots trajectories (to the right only) for initial positions entered from the keyboard. The Frobenius program allows input of the first 5 terms and plots 5 approximations; the Fourier program plots three approximations.

The package is available for $75 from Mathe Graphics Software, 61 Cedar Road, East Northport, NY 11731.

EZQ. This is one of the more powerful products available for the Apple II, but it is not one of the easiest to use. The program will solve first and higher order differential equations and systems of equations. It has very good graphics capability.

Functions are built up by filling in blocks that take the exponential of a number or multiply two results or integrate a result. The program is designed the way an analog computer would be built to solve the problem. Depending on the amount of memory available, 32 to 96 blocks can be used at one time.

The user needs to think about the proper sequence of blocks to solve a given problem. This approach takes some time to learn, so this is probably not a product that one would hand to students and ask them to produce their own solutions. However, a sequence of blocks can be saved on a disk and used later. Therefore, the program could be used by the instructor to work out a problem, save it on disk, and either show it in class or give the disk to the students.

Prices for EZQ range from \$79.95 to \$199.95, depending on the level. It is sold by Acme Software Arts, Box 6126, Evanston, Il 60204.

Differential Equations. This program has versions available for CP/M, Apple II, Atari, Commodore, TRS–80, and MS–DOS computers. It includes routines for Euler, improved Euler, Runge–Kutta for first and higher orders, and Richardson's Extrapolation; the last can be used to improve results from the other methods. The program has character graphics only (printing * is the nearest position), which is rather crude. For some types of computers, however, it might be the only package available for differential equations.

The program is distributed by Dynacorp, 1064 Gravel Rd., Rochester, NY 14618, and it costs \$19.95.

Animation Demonstration by Eric T. Lane, University of Tennessee at Chattanooga. This is not really a differential equations package, but rather a collection of demonstrations of physical laws and properties of wave and particle motion. It is available now for the Apple II, and versions for the Macintosh and for MS–DOS machines are to be released fall 1987 and fall 1988, respectively.

The graphics demonstrations are extremely good. The topics covered include standing and travelling waves, group velocity, Doppler effect, relativistic electromagnetic waves, electron waves, constructive and destructive interference, state transitions, kinetic theory, Brownian motion, Maxwell's demon, electron conduction, and gravitational and magnetic effects.

This is a real buy for a physics lab, and it is of interest for a course in differential equations. The price is \$25, and it is available from CONDUIT (address above).

Math 246 Programs by C. H. Cook, Garry Helzer, and James A. Hummel, all at the University of Maryland. This package requires an IBM PC or compatible with math coprocessor, mouse, and CGA compatible graphics card. The package consists of a menu driven selection of 5 programs: *direction fields*, *separated variables*, *damping*, *systems of equations*, and *phase plane*.

Direction fields allows a choice from a menu of five differential equations and plots the direction field. The user can then sketch a trajectory with the mouse and have the real trajectory plotted through a chosen point. *Damping* allows the user to pick values for the damping and the magnitude and frequency of a forcing function for a vibrating spring, which is shown while the graph is drawn. *Systems* solves systems of the form:

$$\frac{dx}{dt} = ax + by, \frac{dy}{dt} = cx + dy.$$

The eigenvalues are computed, and the tangent fields and trajectories are drawn. *Phase plane* allows the user to pick from a list of differential equations and plots the x–t, y–t, and x–y graphs.

This is an excellent package, and the price is right. You can get it by sending a disk to Garry Helzer, Department of Mathematics, University of Maryland, College Park, MD 20742.

ODE Demonstrations by J. M. A. Danby, North Carolina State University. The origin and uses of these programs are described more fully in Tony Danby's article in this volume. There are versions of the package available for Apple II and for MS–DOS computers with a math coprocessor. The disk contains 50 programs, and they are not menu driven. However, most of them are easy to use. The programs include tangent field with trajectories, Euler and modified Euler method, predator–prey, loan balance, chemical mixing, damped oscillation, and many more applications. The instructions are well written. There are some programs that surprise the user with Danby's humor.

Again, the price is right for software to illustrate the applications of differential equations: Send a disk (two disks if you want compiled versions of the MS–DOS programs) to the author at Department of Mathematics, North Carolina State University, Raleigh, NC 27695.

Spring and String by Howard Lewis Penn, U. S. Naval Academy. This package runs on MS–DOS machines with graphics card. It consists of specialized programs for two applications of differential equations. This review cannot be considered unbiased, of course.

Both programs have very good graphics. The spring program allows the input of values for the mass, damping constant, spring constant, magnitude and period of a sine wave forcing function, and initial position and velocity. The program displays the analytic solution and shows the spring moving up and down while the graph is traced out.

The string program has three portions. The first displays the partial sums of the Fourier series solution for one of five problems at a specified time. The second displays the Fourier series solution at various times; the user controls the number of terms used and the number of points displayed. The third shows the d'Alembert solution.

The package is free; send a disk to the author at the Mathematics Department, U.S. Naval Academy, Annapolis, MD 21402.

REFERENCES

1. Bridger, M. "Review of *Differential and Difference Equations through Computer Experiments*." **BYTE 11**, No. 7 (July, 1986), 63–66.

2. Cunningham, R. S., and D. A. Smith. "A Mathematics Software Database." **The College Mathematics Journal 17** (1986), 255–266.

3. Cunningham, R. S., and D. A. Smith. "A Mathematics Software Database Update." **The College Mathematics Journal 18** (1987), 242–247.

4. Riegsecker, J. "Review of *Elementary Numerical Techniques for Ordinary Differential Equations*." **The College Mathematics Journal 17** (1986), 182.

Computer Applications in Differential Equations

J. M. A. Danby

The course considered in this article is typically called "Introduction to Differential Equations with Applications." It is taken mostly by fourth semester students with majors in engineering, physics, or mathematics, and it is concerned only with ordinary differential equations. It is usually a *service* course, so that the rationale for its existence lies in the applicability of differential equations, not in the theory.

The contents and pedagogy for such a course are hallowed by tradition and have endured for many years, as can be seen by comparing old textbooks with those now appearing; the evolution has been largely cosmetic. In spite of the ingenuity of several authors, the *applications* have changed very little. The reason for this is obvious. The equations that are allowed into an introductory course must be solvable by students in an examination. Physical situations associated with such equations are so restricted that few of them are scientifically non–trivial, and most are not nearly as interesting as we might think. We would surely like to demonstrate, in our teaching, that differential equations are useful and often essential in the world outside the classroom; but too often we just become entangled with scientific triviality and the pedagogy of the "word problem."

Currently, it is the differential equations that seek out the applications — a situation diametrically opposed to the priorities of science. Fortunately, we can now break with tradition. Many physical situations that are familiar from our own and from our students' experiences can be modelled reasonably and simply using differential equations; but these equations cannot be solved by formula in an examination. However, the equations can be set up, varied, discussed, and solved in the classroom, if a computer is used. Outside the classroom, in (dare I say it?) the "real world," most applications will be investigated in precisely this way.

Shortly, I shall describe some of the applications that I have used in class and assigned to students. But first, a digression is in order on the subject of "the computer." The phrase elicits diverse reactions. Some think of the computer in terms of "programming," with jealous bias for (or against) this language or that. Programming is a repellent activity to many students who have been conditioned to regard it as a ritual in which they are constantly risking the wrath of the gods of computing (i.e., instructors in computer science) for even minor infractions of style. If a computation is correct, the output will be the same, whether the program was one line of APL or pages of Italianate BASIC. In fact, many simple, unpretentious, but effective programs can be prepared for class use with no more trouble than it takes to prepare lecture notes. For some, the opportunities and limitations of the computer are embodied in the available "software" — a pity. Other people's programs can be a source of ideas and inspiration, but they should not put bounds on our own use of the computer.

For me, by far the most important attribute of the computer is that *computation* is now easy, and with respectable precision. This fact might prompt us to consider the differential equations course through the eyes of a mathematician of the early nineteenth century (Gauss, for instance), when the urge to compute was the norm, and when much mathematics was developed constructively to solve problems by computation. When the art of computation later remained stagnant, it was abandoned by most mathematicians and omitted from mathematical instruction. The situation has now changed totally, and this alone should make us review what and how we teach.

Apart from ease of computation, another major freedom granted by the computer is *graphics*. In the classroom this is perhaps the most exciting freedom of all, as it can be used to bring subject matter to life. Traditionally, differential equations have been presented as *static* entities with a string of methods for their solution. This should stop. A differential equation is like a machine, capable of movement and ready for action, which only requires some activation: pressing a button, inserting a coin — or (when using a computer) entering a run command.

Demonstrating the fundamentally *dynamic* character of differential equations is educational and can help sustain interest in class. Interaction with a computer is easy, so students can choose parameters, vary initial conditions, and see the consequences immediately. Any application can be made more interesting. For example, it can be salutary (if depressing) to see what is actually happening when a loan is being repaid, as a function of the interest charged and the time allotted for repayment.

Direction fields, with solutions superposed, can be shown routinely. Phase plane diagrams can be introduced and plotted. Stability characteristics can be illustrated. The solution of an equation can be brought to life instead of being left as a static and often arcane formula.

The Laplace transform is often included in the traditional course, but students may be unwilling to accept it as anything but a nuisance when it is used to solve equations that were also solved by more elementary methods a short while earlier. It is a different matter when equations with discontinuous forcing terms are introduced, but many students have real problems in using the step function. Graphics can help: The step function can be programmed, computed, and plotted, to show that it really is a bona fide function. It can be combined with other functions and the results displayed, so the difference between, say, $u(t)sin(t)$ and $u(t)sin(t-1)$ will become easier to appreciate. For differential equations with discontinuous forcing terms, I display simultaneously the forcing term and the solution, both for problems worked in class and for some homework problems. The delta function can be made more dramatic by making the computer squeak when the function is applied! The computer can be used as a medium to make mathematics seem less abstract; one should not ignore the beneficial effects of using it with a humorous touch.

I have heard it argued that since computation is necessarily approximate, the use of the computer runs contrary to mathematical rigor. Actually I find that existence and uniqueness questions arise naturally, and even urgently, when computation is involved. Before starting to compute, it is well, to put it mildly, to make sure that the computer will be operating in a proper domain. It is easy to demonstrate to students the kind of nonsense produced when a computer *operator* acts stupidly; then one can point out how the operator could have avoided being fired, if only he had used the appropriate theorem.

In an introductory course it is virtually impossible to include a balanced survey of numerical methods for solving ordinary differential equations, and there is no consensus on how the topic should be included. Should there be a general survey of methods? Should esoteric (if vitally important) topics such as stability be included? Should the students be voyeurs while methods are described and demonstrated, or should they perform calculations themselves? If the students are to compute, what computing facilities are needed? Should the topic be treated separately or integrated into the coursework? Should just one method be taught and used throughout the course to solve problems?

At present, computing facilities and the students' experiences are so varied that it is hard to require every sophomore to compute except in a course designed to teach computing. Also, many of the faculty are unready or unwilling to see the computer in the classroom. At North Carolina State University the issues of teaching numerical methods and more general applications are addressed in part by offering a supplementary course (a kind of laboratory) that is voluntary, meets for one hour a week, and carries one hour of credit. Students use the University's computing facilities or their own computers. A little programming experience is needed, but the choice of programming language is up to the student.

The fundamentals of stepsize, local truncation error, and order of a method are taught first, with computing assignments. Euler's method is used to illustrate these, but I stress that this is not a viable method in this day and age (a fact that could be mentioned in more textbooks). Then Fehlberg's 4(5) Runge–Kutta method with stepsize control is introduced. This control is essential in many applications (an orbit from the Earth to the Moon, for instance), and with Fehlberg's method it is remarkably easy to understand and to use. Coded programs are available to the students, and they perform assignments to learn how (and how well) the method works. Students work with various differential equations, and they are encouraged to communicate with one another. (Students don't do that enough; computing together generates dialog.)

Each student chooses two projects from a list of eighty in the text, ranging from the physical and biological sciences to economics. These are mostly topics that a student traditionally might meet only in an advanced course, if at all. Today they are easily accessible to sophomores. While working on two projects, a student will see others illustrated and discussed in class, so will gain experience with a wide range of applications. The grade for the course depends on the timely performance of the projects; there is no written examination. A course of this kind generates a lot of contact and discussion with individual students and is very rewarding to teach.

I give here short descriptions of some of the projects that I have used in class. Considered as student projects, they do not require computers with graphics capabilities. For demonstration in class, it is essential to use graphics;

for most projects, it is almost essential to use the additional speed of the 8087 math coprocessor chip.

The first topic is that of "fireworks." Actually this is not suitable for student projects, but it is excellent for demonstration and discussion. Every particle ejected from a firework moves subject to the same differential equations. The initial conditions of the particles determine the type of firework. The equations describe three–dimensional motion in the (constant) gravitational field of the Earth, without resistance, or, better, with resistance proportional to the velocity. For one of the best known fireworks, the particles are ejected at the same time from the same position with the same speed but in random directions. Here the equations can be solved by formula, and one can show that at any instant, the particles lie on the surface of a sphere. With resistance, the radius of the sphere tends to a limit; without resistance, the radius is unbounded. The viewer, whose position the programmer can choose, will see the three–dimensional display as if projected onto a plane (the screen). Positions are plotted at equal increments of time; I find it most effective to plot a new position as soon as it is found to help the impression of constant movement. This produces a gratifying display, which can be strengthened by having more than one firework simultaneously on the screen, in different places, starting at different times, and in different colors. All this has the serious aspect of showing the effects of changing initial conditions. (One way to display this model is to show current positions of particles but erase previous positions; then the expanding sphere is observed.) It is easy to think of other firework models.

The spread of disease is, perforce, a subject of considerable interest to today's student. The model often found in texts has a fixed population N, which includes S that are susceptible to a disease and I that are infected and infectious. Then

$$\frac{dS}{dt} = -aSI,$$

$$S + I = N.$$

This is easy to solve but quite unrealistic. (However, you can have fun in class if N represents the student population, and the disease is carefully left unspecified.) A more realistic model includes the number R who are "removed," i.e., isolated for treatment, dead, or cured and immune. Then

$$\frac{dS}{dt} = -aSI,$$

$$\frac{dI}{dt} = aSI - R,$$

$$\frac{dR}{dt} = bI.$$

Since $S + I + R = N$, these can be reduced to two equations. For constant coefficients, these can be solved if the time t is eliminated. But the solution,

$$I = I_0 - (S - S_0) + \frac{b}{a} \ln \frac{S}{S_0},$$

is not, by itself, very informative. More interesting cases arise if a, which measures the infectiousness of the disease, is not constant; for example, there might be seasonal variation. Or the rate of removal may not be linear in I. Or vaccination, with differing tactics, may be introduced. Or the total population N might vary. These changes can be made to a program in a matter of seconds during class and the effects observed.

Another class of models for the spread of disease involves more than one species. Malaria falls in this category, and so does the well–known model for the spread of gonorrhea, which is distinctive in that there is no immunity. Here, if x out of m males and y out of f females are infected, then

$$\frac{dx}{dt} = -ax + by(m - x),$$

$$\frac{dy}{dt} = -cy + dx(f - y).$$

A model of this kind can be discussed partly in terms of equilibria and stability. Again, the coefficients can be varied, and m and f can be varied with the time. I include a model for the spread of the AIDS virus among my projects; although I have not yet discussed this in class, several students have chosen to work on it.

Predator–prey models can be constructed in great variety using the simplest of reasoning. That of

Volterra, which is justly the most famous, has the equations

$$\frac{dx}{dt} = ax - bxy,$$
$$\frac{dy}{dt} = -cy + dxy,$$

where x and y are the populations of the prey and predator, respectively. A great merit of this model is that it produces the cyclic behavior often observed in nature.

In the population growth of one species, exponential growth is usually seen to be unrealistic, logistic growth being superior. If the prey population is limited by environmental factors as well as predation, this suggests modifying the basic model to

$$\frac{dx}{dt} = ax - bx^2 - cxy,$$
$$\frac{dx}{dt} = -dy + exy.$$

This, it turns out, results in the loss of the cycles. Logistic growth can also be introduced among the predators. If

$$\frac{dx}{dt} = ax - bx^2 - cxy,$$
$$\frac{dy}{dt} = y(d - ey/x),$$

then the predator population is limited by the food supply, i.e., the numbers of the prey. This model is also unsatisfactory if we are looking for cycles.

The model for predation is unrealistic; in practice a predator will eat as much as he needs and then stop. This implies that, as x becomes large, the rate of predation should be proportional to y. A model satisfying this requirement is

$$\frac{dx}{dt} = ax - bx^2 - \frac{cxy}{x + d},$$
$$\frac{dy}{dt} = y(e - fx/y).$$

This model can have a stable limit cycle, while the usual equilibrium is unstable. All of these models, and more, can be discussed in class. It takes only a few seconds to change a line in the subroutine or procedure that defines the model, and it is all the more convincing if the students actually see this done.

Limit cycles are usually avoided in an introductory course because of the theory involved; but they happen all around us. A good example is the motion of the pendulum of a clock. Kinetic energy is boosted by a fixed amount at the bottom of each swing; if a resisting term proportional to the angular velocity of the pendulum is included in the model, then the limit cycle can be seen clearly. If the initial swing is too large or too small, the limit cycle is soon approached. This can be illustrated by a phase–plane diagram; you can even make the computer produce the appropriate "clicks." This model is unsatisfactory in one respect: The limit cycle is approached, however small the initial swing. From practice, we know that with a small enough initial swing, the pendulum will stop altogether. This can be modeled by adding "dry friction" to the model. Dry friction can occur when two solid materials are in contact; the resisting force is greatest when the relative speed is least. This is very common in nature, and it is a good idea to introduce it in our applications. The contact between a violin bow and string, for example, is subject to dry friction.

In a dynamical system such as a pendulum, the motion can often be discussed conveniently in terms of parameters, in this case amplitude and phase angle. Physical insight can be gained by following the variation in these parameters, hence, the method of variation of parameters. In our courses this method is applied only to linear equations; it is also attributed to Lagrange. In fact, the method goes back to Newton, and its useful applications in mechanics have been to nonlinear systems, where the resulting equations require numerical integration. If we are to teach the method of variation of parameters, we can now do so objectively.

Students enjoy orbits. This is reflected in the frequent discussion of planetary motion in calculus and in some differential equation texts. (I have yet to see such a discussion that is completely correct!) Most orbital problems require numerical integration. An entertaining project is to follow the descent of "Skylab" through its final few revolutions. In this model atmospheric resistance is added to the inverse square gravitational field of the Earth. The same model can also be used for projects involving ICBM's.

Another simple model follows orbits in the fields of the Earth and Moon. It is sufficient to place the Moon in a circular orbit around the Earth and to follow motion in just two dimensions. A spacecraft starts from a parking orbit around the Earth; the radius of this orbit, and the position and speed of launch can be specified. The object is to hit the Moon, but many other interesting orbits can occur. An orbit might loop around the Moon and return to the Earth. Or a passage close to the Moon can result in the spacecraft receiving a boost in energy; this latter type of maneuver has often been used in practice.

A beautiful model that is hard to set up, but worth the effort, involves the passage of two "galaxies." Before the passage, one of the galaxies has rings of "stars" moving around it in circular orbits. Following the passage, some stars may have been captured by the other galaxy, and temporary bridge–like structures may join the galaxies. Some such structures have actually been observed between nearby galaxies. In this model the stars move subject to the gravitational field of the galaxies, but they exert no forces on each other or on the galaxies. This model requires stepsize control separately for each star.

For demonstration purposes, Fehlberg's method is too slow, and I use power series solutions. Such solutions are efficient for systems of nonlinear equations; their drawback is that the program must be set up separately for each model. But if the same equations are to be solved many times (as is the case for the galaxies), then the effort is worthwhile. This use of power series should at least be mentioned to students, but it has made little impact on the textbooks as yet.

This project can be interpreted in other ways. The two "galaxies" can become the Sun and Jupiter, and the "stars" will become minor planets. Or the "stars" can become particles in a ring system moving subject the parent planet and a satellite. As another interpretation, the "galaxies" become the Sun and its problematical companion star, "Nemesis," and the "stars" become members of the Oort cloud of comets; a passage of Nemesis can result in the changing of some cometary orbits so they approach close to the Sun.

Many comets approach the Sun from great distances in almost parabolic orbits. But there are many comets with elliptic orbits, and many of these have aphelion distances approximately equal to the distance of Jupiter from the Sun. In a model that is popular with students a comet enters the solar system from a great distance and moves close to Jupiter; if conditions are just right, the passage with Jupiter results in the orbit being changed to an ellipse of the sort just mentioned.

If "computing" and "modelling" are mentioned in the same sentence, then it is natural to think of *difference equations*; indeed, many texts on differential equations include a section on difference equations. A case can be made for further coverage. Some situations that are usually modelled by differential equations make more sense when treated with difference equations. Some features, such as time delay, cannot be considered in a first course on differential equations but are simple to include in difference equations. Random processes may easily be included in a model based on difference equations. Numerical solutions of difference equations are easy to compute on the most modest programmable calculator. An elementary introduction can be given to systems that become chaotic when a parameter is increased. Such systems are of great importance today, and the earlier students hear about them the better. (Indeed, Sandefur and Vogt, in another paper in this volume, propose introducing these systems to freshmen.) Chaotic behavior applies also to differential equations, but is not so easy to demonstrate.

The expedient of offering an additional course to cover computing and applications is unlikely to have wide appeal, so the question arises as to how some of the material might be incorporated into the confines of the traditional course. This would not be hard to do if some of the traditional content were to be dropped; I shall end by suggesting some topics whose omission would harm only traditionalists.

The regular calculus curriculum includes examples of separable differential equations, of exponential growth and decay, and of exactness; we should not repeat this material. Exact differential equations are unlikely (an understatement) to be found outside the classroom, anyway. Most applications of first order equations are boring and useless, with orthogonal trajectories narrowly in the lead. Banks do not use differential equations to compute mortgage payments, so why should we? The Wronskian is nice for proving some theorems, but it is never truly *needed* to test two or three functions for linear dependence or independence. Variation of parameters, as we currently teach it, should go. Power series solutions of linear differential equations will never be used by the majority of students; if they are to be used in a *second* course to discover a special function or two, let them be developed (only in enough detail as is really necessary) at that stage. The extent to which the Laplace transform is included depends on the service nature of a course; but the heavy burden of algebra, which sinks some students without trace, will surely be lightened soon when symbolic manipulation becomes an everyday activity.

We are all guilty of "student bothering," i.e., activity that keeps students busy doing problems that are easy to grade. This may strengthen their characters, but it teaches very little. The curriculum is not all that we need to change, but it will do for a start.

RESOURCES

The text used at North Carolina State University in the "laboratory" section is *Computing Applications to Differential Equations*, by J. M. A. Danby, Reston Publishing Co., Reston, Virginia, 1985. This includes discussion of the models mentioned here, with attributions and references.

Demonstration programs that can be run on an IBM PC with 8087 math coprocessor can be supplied by the author. They are available in two versions: compiled programs on two disks or source code in Turbo Pascal on one disk. Send one disk or two to the address given elsewhere in this volume.

Additional software resources for differential equations are described in the article by Howard Penn in this volume.

Computer Use in Teaching Statistics

Florence S. Gordon

Statistics is fast becoming the single most important mathematics offering for the non–mathematics, science or engineering major. Several prominent groups concerned with mathematics education have called for statistical literacy to be an essential component of a modern college education. This reflects the fact that statistics now pervades almost all facets of our lives.

Most of the tremendous growth in statistics and its applications over the last 30 years (in fields such as business, health sciences, economics, psychology, political science, and educational research) is directly attributable to the availability of high speed computers. As a consequence, we might expect to find extensive use of computers in statistics courses; however, this has been slow to materialize to any great extent. The delay may be due to the almost universal use of hand–held calculators in such courses. Only recently has there been any growth in the use of computers by statistics students to perform computations with commercial packages, especially in a mainframe computer environment. Various publishers have recently released student-oriented microcomputer packages for elementary and applied statistics. The availability of such materials is providing a major impetus to increase the use of computers in statistics courses. The recent CBMS study [by D. Albers, R. Anderson, and D. Loftsgaarden, published by MAA] of undergraduate mathematics indicates that 29% of the elementary statistics courses offered in mathematics departments include some computer component.

I will give an overview of how the computer can and should be used in statistics offerings at the introductory level for a general audience or a specific discipline, as well as in a follow–up course in applied statistics.

The typical non–calculus–based statistics course emphasizes the *concepts* and *methods* of inferential statistics, while the calculus–based introductory statistics course typically emphasizes the *concepts* and the *theory*. Neither places heavy emphasis on the *computations*. Despite this, the first generation of software for statistics education was computationally oriented and, as such, was inappropriate for these courses. At the elementary level, the result of using such software is to have the computer serve as a glorified hand–held calculator. It is certainly desirable to eliminate most of the drudgery associated with doing the computations by hand. However, this is merely one small facet of the enormous potential for computer use in statistics education.

Modern microcomputers provide a powerful tool to transform most statistical concepts and methods into visual representations which can dramatically improve student comprehension. Most business–oriented software packages (for example, *Lotus 1-2-3*, *Symphony*, *SuperCalc*) can display data in visual form via histograms, bar charts and pie charts, and such packages can be used in an elementary statistics course. However, data display represents little more than the first week of the standard introductory course. Introductory statistics (both calculus– and non–calculus–based) can use the power of the computer for simulations and demonstrations of most of the important concepts and methods of inferential statistics, with the goal of increasing student understanding.

The most important aspect of statistics is dealing with data, and it is here that the computer can be most valuable. We should not be content with merely producing THE mean, THE standard deviation or THE frequency distribution of a set of numbers (whether by hand, calculator or computer). Rather, the computer enables us to explore the effects of the data on these quantities because it can *store* as well as *compute*. Students should learn to appreciate how a change in the data alters the mean or the median. They should see how a single outlier can drastically affect the standard deviation. Moreover, given a set of data, students should be encouraged to experiment with the changes that occur in a frequency distribution or in the shape of a histogram or frequency polygon if the number of intervals used or the range of selected values is altered. In this way, we can lead the students to a deeper understanding of the significance and validity of their statistical results.

A new emphasis in statistics is called Exploratory Data Analysis (EDA). It involves exploration on data in a dynamic setting whose effectiveness essentially depends on the use of computers. Some of the available software packages are accompanied by databases of appropriate populations which provide the opportunity for this type of exploration. These databases are also valuable in many other applications that involve generating a large random sample.

The one concept that pervades all of statistics is that of randomness. It is a particularly deceptive topic in the sense that few students truly grasp it. This problem is not always evident in typical classroom situations, but it becomes painfully obvious if students are required to collect and analyze random data. They have no feel for what it means to make a random selection. The use of a table of random numbers seems little more than an unnecessary complication to them.

The computer provides a major assist in conveying some feeling for randomness in statistics. At a simplistic level, it can be used to generate repeated random selections from a set of pre–selected numbers or of Heads and Tails to simulate coin tossing experiments. However, students can achieve a much more effective visual demonstration of random phenomena through computer graphics.

One of the simplest and most accessible examples of randomness is a two–dimensional random walk. The computer can draw the path traversed for any choice of step size and any number of steps. Students are often stunned to see the completely different paths that result from the same choices of step length and number of steps. Their reaction clearly underscores their lack of comprehension of random phenomena.

The concept of randomness can also be approached through a binomial distribution ($p = 1/2$), which is simulated effectively by a coin–tossing experiment. Typically, the user chooses the number N of fair coins, and the program randomly "tosses" sets of N coins, displays graphically the sets of H's and T's, and totals the numbers of heads during repeated trials. This gives the student a better understanding than anything that can be done on the blackboard or with numerical displays alone. First, it gives a visual feel for randomness in terms of the different results obtained on successive runs for the same N and also from an examination of the occasional outlier that arises. In addition, the simulation gives a feel for the shape of the binomial distribution and for the results of changing the number of coins. Moreover, the student sees a convincing numerical check on the accuracy of theoretical predictions. Thus, students can verify that the experimental results are "close" to those predicted and yet see that the expected values are almost never precisely achieved in practice. Such simulations also enable more general binomial experiments with unfair coins, dice rolling, or other random processes. Some graphics programs (e.g., those illustrated in the article by Snell and Finn in this volume) demonstrate effectively the limiting behavior of binomial processes, leading to a simulation of the Law of Large Numbers.

In a different direction, normal probabilities can be displayed graphically by drawing the normal distribution curve for any choice of the parameters and shading and computing the area between any two x values selected by the user. This provides the student a greater appreciation of the relationship between the probability that X falls between two values and the area of the corresponding region under the curve; the numbers in the standard normal tables now have a visual significance. Alternatively, it is possible to simulate areas under the normal curve and the corresponding probabilities by generating normally distributed values, plotting them, and keeping track of the proportion which fall in the indicated region.

One of the important applications of the normal distribution in introductory statistics is its use to approximate the binomial distribution, and this can be treated extremely well using computers. In particular, the binomial distribution can be displayed graphically as a histogram for any choice of probability p of success and number N of trials. The theoretical normal distribution can then be superimposed over the binomial histogram to provide a better understanding of why we can replace the binomial distribution with a corresponding normal distribution and why we need to introduce a continuity correction in such situations.

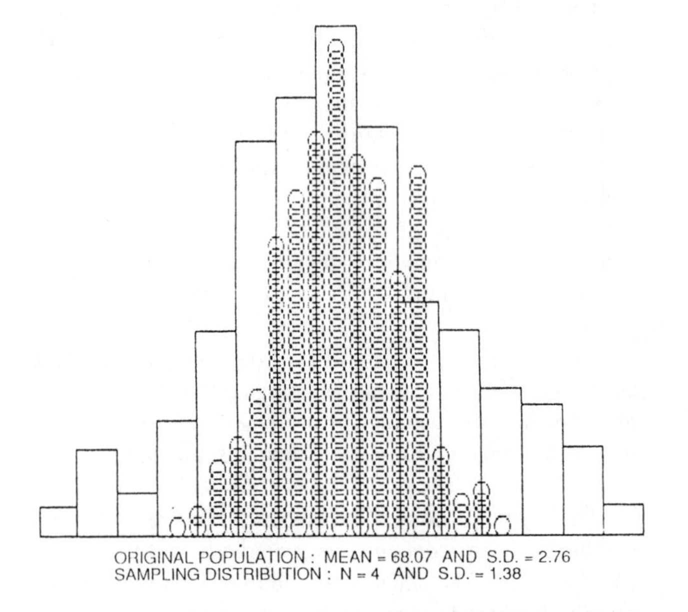

Figure 1. Sampling from a normal population.

The Central Limit Theorem, the primary mathematical tool of inferential statistics, is undeniably the least understood topic in any introductory statistics course. A typical computer simulation of it gives the user a choice of various underlying populations and a desired sample size. It first draws the histogram of the chosen population, then randomly generates the samples, calculates the mean of each sample, and plots it. The resulting distribution of sample means literally grows in front of the user. Figure 1 shows the output from such a program with samples of size 4 drawn from a normal population; Figure 2 shows the output with samples of size 9 from a skewed population.

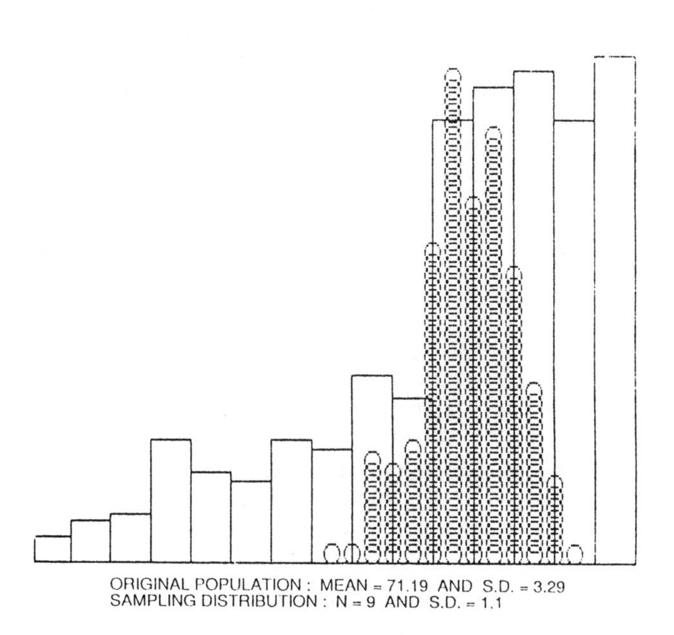

Figure 2. Sampling from a skewed population.

Several successive runs with the same underlying population and successively larger values for the sample size demonstrate clearly the convergence of the sampling distribution to the normal distribution. The graphic displays also show the effects of sample size on the standard deviation. A rough visual estimate of the spread of the sampling distribution can be compared to the spread of the underlying population to indicate approximate proportions. Many students actually anticipate the formula for the standard error of the sample mean from this visual analysis. When the displayed values for sample mean and standard deviation are compared to the theoretical values predicted by the Central Limit Theorem for the distribution of sample means, the theory is effectively validated.

Another effective use of computer graphics in a classroom is to demonstrate convergence of the t-distribution to the normal distribution as the sample size increases. The graphs of the t-distribution for several values of N (between 2 and 30, say) are drawn, and the standard normal curve is superimposed. From this, the convergence of the t-distribution to the normal becomes visually clear to the students. (See Figure 3.)

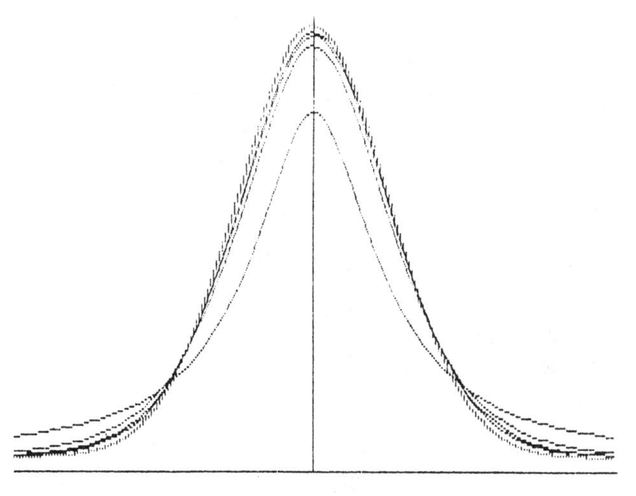

Figure 3. Convergence of t-distribution to normal. (The normal curve is highest at the center.)

Estimation via confidence intervals and hypothesis testing are other standard topics in elementary statistics which students often treat mechanically with little comprehension. An appropriate program can do much to improve the level of understanding. For instance, the user can be led through the preliminary analysis in an estimation problem: Is the confidence interval for mean or proportions? Is the sample size large or small? The program then requests the appropriate data: mean, standard deviation, sample size, and confidence level (for means), or sample proportion, sample size, and confidence level (for proportions). The output might be the numerical confidence interval based on the data and/or a graphical display showing the locations of all the usual confidence levels. This gives greater insight into the validity of the estimation process. It also enables exploration of the effects that sample size or changes in the data have on the confidence interval and hence on the prediction.

The solution technique for a hypothesis test problem is similarly structured as a series of sequential steps and decisions, including statement of the null and alternate hypotheses and selection of one or two tails, type of distribution, and critical values. Many students blithely skip over much of this analysis. A computer program forces them to focus on the necessary steps and develops the corresponding methodology in their work habits. These topics are also suitable for effective simulations that demonstrate the effects of different random samples on either the confidence interval or the acceptance or rejection of a null hypothesis.

Regression analysis lends itself to computer use via a program to draw the scatterplot for any set of input data and the corresponding regression line. The speed of the computer allows the student to see quickly the effects on the resulting regression line of changing, adding or deleting a point, thereby getting a better understanding of the concept. This allows the student to turn regression analysis into an experimental activity. The student is also able to see the effects of altering the data on the values of the regression coefficients, the correlation coefficient, and the standard error of the estimate.

In summary, the computer provides the opportunity for an elementary statistics instructor to direct the emphasis away from a collection of standardized methods the student must apply mechanically and toward an understanding of the underlying statistical concepts. Hopefully this will give students a more lasting appreciation of statistics and will help achieve the goal of universal statistical literacy among college graduates.

A follow–up course builds on the fundamental ideas studied in an introductory course and quickly goes into a variety of more advanced and sophisticated statistical topics, such as linear and nonlinear regression, correlation, cross–tabulation, analysis of variance and covariance, the chi–square distribution and contingency tables, goodness of fit, time–series analysis, factor analysis, and non–parametric tests.

Most of these topics entail an enormous amount of computation, which makes them inaccessible for hand or calculator treatment. Therefore, sophisticated computer packages are essential and have long been available to perform the required number crunching. These packages, such as SAS (Statistical Analysis System), SPSS (Statistical Package for the Social Sciences), Minitab, and BMD (Bio–Medical Package), were designed primarily for a mainframe computer environment; some (e.g., SAS and SPSS) are now available on microcomputers. They are essential tools for managing the large quantities of data that often arise in real–world problems and, as such, are especially useful to the applied statistician. Because they were designed for professional use, they tend not to be particularly user–friendly and so can be both inappropriate and intimidating for students. By their very nature, such packages presuppose a knowledge of the statistics upon which they are based rather than being intended for learning the statistical tests.

Recently a number of statistical packages for microcomputers have been published which are intended for educational use. They tend to be more user–friendly and are usually accompanied by textbooks rather than professional manuals. A student using such a package does not have to be concerned with the computational drudgery of the statistical procedures, but can concentrate on the significance of the statistical analysis and so better appreciate the applicability of the procedures.

Beyond the use of the computer at this level as a purely computational tool, there have been some attempts to use computer graphics for displays of multivariate data and concepts of multivariate analysis. Especially in three dimensions, the visual display provided by the computer is extraordinarily effective. For example, it is possible to graph both the observed and expected frequencies in a chi–square analysis problem to "see" whether there is an apparent significant relationship between the factors. This type of insight is certainly a desirable feature to develop in a student.

SOFTWARE RESOURCES

Professional Statistical Packages

1. *Biomedical Data Processing* (BMD), BMDP Statistical Software, Inc., 1964 Westwood Blvd., Suite 202, Los Angeles, CA 90025.

2. *Minitab*, Minitab, Inc., 215 Pond Laboratory, University Park, PA 16802.

3. *Statistical Analysis System* (SAS), SAS Institute, Inc., Box 8000, Cary, NC 27511.

4. *Statistical Package for the Social Sciences* (SPSS), McGraw–Hill Publishing Co., 1221 Avenue of the Americas, New York, NY 10020.

Instructional Packages

5. Doane, D. P. *Exploring Statistics with the IBM PC.* Reading, MA: Addison–Wesley, 1985 (text and disk).

6. Dowdy, S. *Statistical Experiments for BASIC.* Boston: Prindle, Weber, and Schmidt, 1986 (text and disk).

7. Elzey, F. F. *Introductory Statistics: A Microcomputer Approach.* Monterey, CA: Brooks/Cole, 1985 (text with Apple II disk).

8. Frankenberger, W., and T. Blakemore. *Introductory Statistics Software Package.* Reading, MA: Addison–Wesley, 1985 (includes non–graphic Apple II disk).

9. Kemeny, J. G., T. E. Kurtz, and J. L. Snell. *Computing for a Course in Finite Mathematics.* Reading, MA: Addison–Wesley, 1985 (includes True BASIC disk).

10. Lyczak, R. A. *Elementary Programming for Statistics.* Boston: Prindle, Weber, and Schmidt, 1980 (non–graphic FORTRAN and BASIC programming).

11. Presby, L. *COMPSTAT: Solving Statistical Problems by Microcomputer.* New York: Random House, 1984 (includes non–graphic Apple II and IBM PC disks).

12. Triola, M. F., and W. J. Flynn. *STATDISK.* Menlo Park, CA: Benjamin/Cummings, 1986 (includes non–graphic Apple II and IBM PC disks).

13. Weissglass, J., N. Thies, and W. Frazier. *Hands On Statistics.* Belmont, CA: Wadsworth, 1986 (includes Apple II disk).

The Use of the Computer in a Probability Course

J. Laurie Snell and John Finn

The idea of using the computer in the teaching of probability has been beautifully motivated by Feller in the second edition of his famous book *Introduction to Probability Theory and its Applications* [1]. In the preface Feller writes about the fluctuation of one's fortune in coin tossing:

> These results are so amazing and so at variance with common intuition that even sophisticated colleagues doubted that coins actually misbehave as theory predicts. The record of a simulated experiment is therefore included.

The simulation Feller included was the result of playing a game of "heads or tails" 10,000 times. Today, we can easily ask our students to write a program to carry out this experiment on their favorite microcomputer. At Dartmouth they would do this on their Macintosh using either Pascal or BASIC. Those using BASIC would use a new form of structured BASIC called True BASIC, and their program might be as below.

Figure 1 shows a sample output from this program (with labels added).

Running this program a few times, students learn that being behind or being ahead most of the time is rather typical, and they are more philosophical if they find themselves continually behind in their next poker game.

Peter's fortune is like a random walk in one dimension: Whether Peter will come back to a fortune of 0 is the same as whether the walker will come back to his starting point. The students might be asked to see if they can decide this question by simulation. The more adventuresome instructor will surely mention Polya's beautiful theorem that a random walker in one or two dimensions will return, but not so in three dimensions. Simulations of this problem indicate that we can conjecture results about infinite processes from finite simulations.

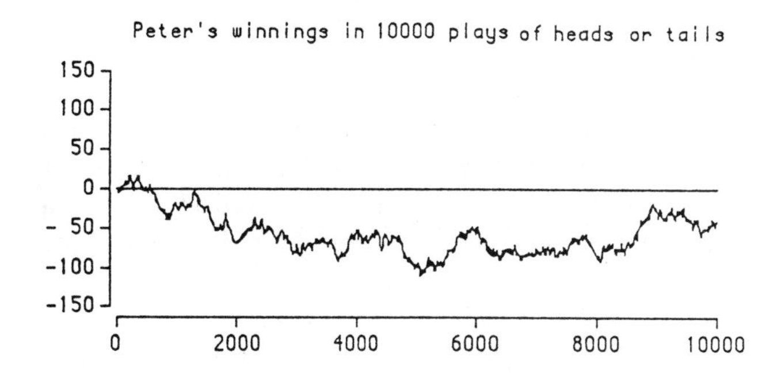

Figure 1. Peter's winnings.

This very first example shows why computing is such an important tool in probability. First, it carries out

```
!Plots Peter's fortune in 10,000 plays of heads or tails.

set window -150, 10000,-250,250              ! Set window coordinates...
plot 0,  0; 10000,  0                        ! & plot axes.
plot 0, 250; 0, -250
let fortune = 0                              ! Initialize fortune.
for game = 1 to 10000                        ! Play the game:
    if rnd < 1/2 then                        ! Heads you win;
        let fortune = fortune + 1
    else                                     ! Tails you lose.
        let fortune = fortune - 1
    end if
    plot game, fortune;                      ! Plot results.
next game
end
```

something that the student cannot reasonably do. (A student actually *could* toss the coin 10,000 times but would then have little time for anything else.) Second, writing the program is simple, enjoyable, and instructive.

There is still a need for some expert programming to provide classroom demonstration programs that produce nice looking and instructive results. Such programs can provide proper labels and easy inputs to encourage the student to play around with them. Thus, at Dartmouth we provide simple programs as examples and for the students to modify, but we also provide pretty versions for demonstration purposes in and out of the classroom. The simple programs and illustrations we discuss in this paper are taken from [2], and the fancier examples were developed by John Finn, working under a grant from the Keck Foundation. We give students the code for both versions, so they can see that it is not difficult to add the nicely labelled axes and special inputs. The more adventuresome or aesthetically minded student will surely want to do this on occasion to his own programs.

We use simulation early in our probability course to illustrate some of the basic problems to be studied, such as the famous Buffon needle problem that estimates π from an experimental estimate of the probability that a randomly thrown needle will cross a line on a ruled floor (Figure 2).

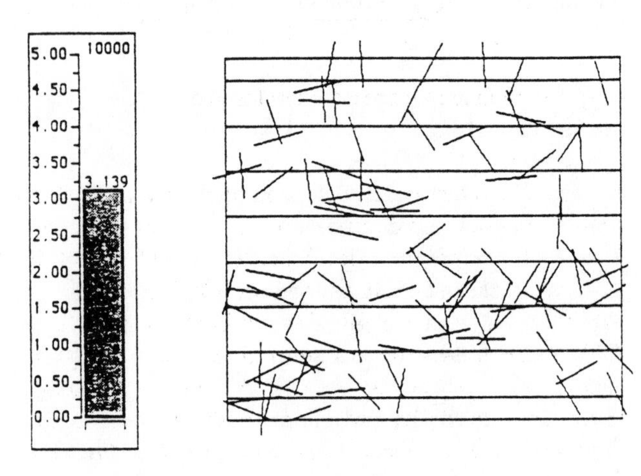

Figure 2. Simulation of the Buffon experiment to estimate π.

This example serves two purposes. First, it shows one of the many ways that probability finds its way into other branches of mathematics. Second, it suggests that a lot of experiments are necessary to get accurate results and raises the question of "how many?". Students will learn later that the error can be expected to be no more than $5/\sqrt{n}$, where n is the number of experiments. Hence for 10,000 experiments it should be no worse than .05 — but that isn't very good!

In demonstration programs like this, we find that display devices, such as the thermometer that keeps a running count of the estimate, both keep the attention of the student and are in themselves instructive in showing the kind of variation that occurs during the simulation. It shows that one could cleverly stop the program at an opportune time, a familiar trick in some ESP experiments.

Computing can help the student understand basic concepts in probability such as conditional probability. The medical students at Dartmouth were recently asked to consider the following problem: A doctor suspects that a patient might have a certain kind of cancer. The incidence of this kind of cancer in the general population, similar in age, sex, etc. to her patient, is one per thousand. A test can be carried out that gives an accurate diagnosis 99 percent of the time when the patient has cancer (test positive) and 95 percent of the time when the patient does not have it (test negative). The test is given and the result is positive. The medical students were asked if they would then think that there was a greater than 50/50 chance that the patient had the cancer. Of course, most said yes, and the lecturer then did the simple calculations to show that the probability that the patient had cancer had, in fact, increased only to about .02. He then pointed out that this surprising result is caused by the fact that the disease was so rare in the first place.

The instructor then asked the same question with the incidence of one per thousand in the general population replaced by five per hundred. Now the students were more divided in their guesses. The lecturer gently chided them for continuing to guess and pointed out that it is very hard to develop an intuition on problems like this. He suggested that they should get used to calculating these probabilities. Few of our medical students will remember how to do this by the time they become doctors, so it is worth showing them that it is a simple matter to write a program to do it. This problem really amounts to calculating two "tree measures," one with time forward and one with time backwards.

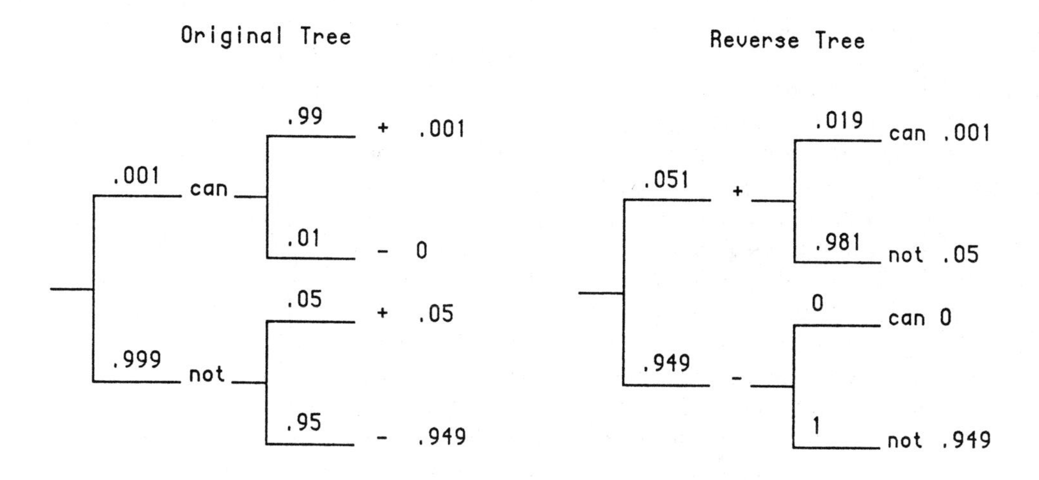

Figure 3. Calculating Bayes probabilities by tree diagrams

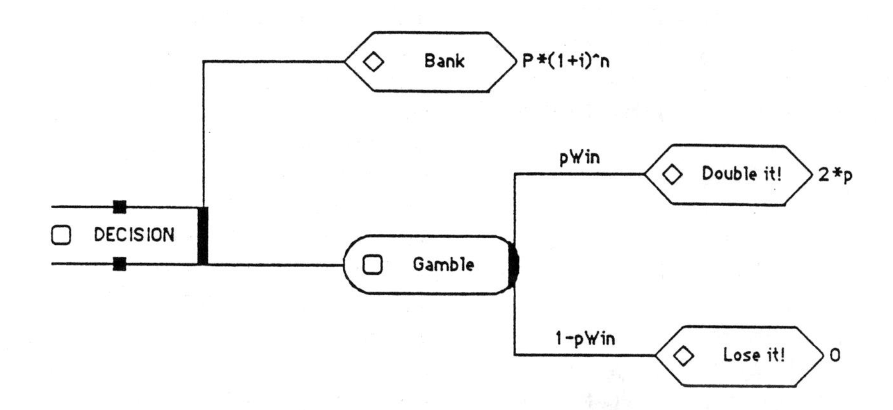

Figure 4. Decision tree for the problem of banking or gambling to increase one's fortune.

In Figure 3 we show a run of a program to carry out these calculations. The run corresponds to the lecturer's first example.

It is natural to extend these tree diagrams to discuss decision analysis. Students can develop programs that analyze simple decision trees, and even ones that perform primitive sensitivity analyses. In fact, a spreadsheet program such as Microsoft's EXCEL can be used by students unfamiliar with programming to accomplish elementary tasks in tree building and analysis.

For more complicated decision problems, a software tool for decision tree construction and analysis is needed. D–MAKER, a program developed for the Macintosh at Dartmouth Medical School by Professor J. Robert Beck and his colleagues, is available from Dartmouth College. D–MAKER allows the student to draw a tree using the mouse, and permits algebraic solution of problems, Bayesian probability revision, multiattribute utility functions, and multiway sensitivity analysis. Figure 4 illustrates its output.

```
! Draws a Pascal Triangle.

dim c(0 to 50,0 to 50)                         ! Square array for values c(n,j).
for n = 0 to 10                                ! For each n..
    for j = 0 to n                             ! and each j..
        if j = 0 or j = n then                 ! calculate c(n,j)...
            let c(n,j) = 1
            let c(n,j) = c(n-1,j-1) + c(n-1,j)
        end if
        print tab(40+3*(-n+2*j)); c(n,j)       ! & print in the right place
    next j
next n
end
```

```
                              1
                           1     1
                        1     2     1
                     1     3     3     1
                  1     4     6     4     1
               1     5    10    10     5     1
            1     6    15    20    15     6     1
         1     7    21    35    35    21     7     1
      1     8    28    56    70    56    28     8     1
   1     9    36    84   126   126    84    36     9     1
1    10    45   120   210   252   210   120    45    10     1
```

Figure 5. Pascal's triangle.

In probability we use a lot of combinatorics and much of this is simple recursion, ideally suited for the computer. For example, the binomial coefficient $c(n,j)$, representing the number of ways that a committee of j people can be made from a group of n people, can be obtained from the famous Pascal triangle. The student who writes a program to construct this triangle will be apt to remember the basic recursion relation that determines the triangle. And the program is again easy. The output from this program is the familiar triangle shown in Figure 5.

This is a great example to illustrate Polya's technique of looking for patterns. Students can be asked to modify their program to print out the results mod 2, i.e., to print a * only when the result is odd, obtaining the output shown in Figure 6.

Can the students see why all entries in the rows $2^n - 1$ are odd? If they can, they may even see the "self similar" aspect of this triangle — a very modern concept in probability. If the students have color monitors, they can see beautiful patterns by printing out the results mod k for other values of k. This is also one of the simplest examples of cellular automata and is often used as an introduction to this topic.

The use of subroutines and libraries allows the student or the instructor to write a sequence of simple programs, each instructive in its own right, and then combine them to handle a complex probabilistic situation. To illustrate this, consider the Central Limit Theorem, a theorem basic to probability but difficult to prove in a introductory course.

The Central Limit Theorem is best discussed first for the case of Bernoulli trials. That is, we have n independent experiments, each of which results in success with probability p and failure with probability $q = 1 - p$. For example, toss a coin n times with "heads" success and "tails" failure. Or try a drug, known to be successful a fraction p of the times, on n patients. Then

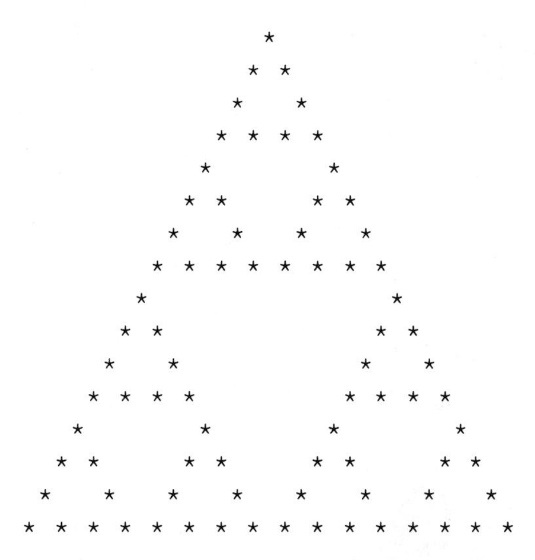

Figure 6. The odd entries in the Pascal triangle.

the probability of exactly j successes is given by $i(n,j)p^n q^{n-j}$, where $c(n,j)$ is the binomial coefficient.

We first define a function that gives these probabilities:

```
function binomial(n,p,j)
    let q = 1-p
    let c = 1
    for k = 1 to j
        let c = c* (n-k+1)/k
    next k
    let binomial = c*p^j*q^(n-j)
end function
```

This function can be used to work problems that require binomial coefficients. It can also be used in a program to plot the binomial density.

We include this program to show how simple using graphics is for the students. Unfortunately, not all languages are as easy to do graphics in as True BASIC. This program also illustrates the use of libraries. The function **binomial** has been put into a collection of subroutines and functions that can be called from any program. Of course, if your language does not have library possibilities, these routines can be pasted into every program where they are needed.

```
! Plots a binomial density

library "Lib.prob"
declare function binomial

input prompt "Probability for success, trials? " : p,n
let q = 1 - p                                      ! Probability of failure.
let max_k =int(p*(n+1))
let height = 5/4*binomial(n,p,max_k)
set window -1/2, 2*max_k+1/2, 0, height            ! Scale the window and
                                                   ! plot axes.
plot -1/2,0; 2*max_k,0
plot 0,height; 0,0
for j = 0 to n
    let b = binomial(n,p,j)
    box lines j - 1/2, j + 1/2, 0, b               ! Draw a bar.
next j
end
```

Our pretty version adds labels and allows the user to input several values of n, as we show in Figure 7.

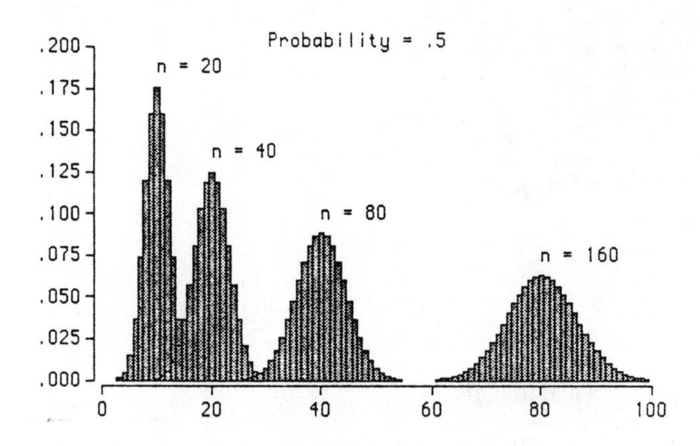

Figure 7. Binomial densities.

When these binomial densities are demonstrated in class, it is natural to observe that the graphs tend to flatten out, drift off to the right, and become bell shaped. This motivates the Central Limit Theorem and the normalization used in this theorem. Having stated the theorem, a modest change in the program **Binomial density** provides the final normalized densities and compares them with the normal curve as in Figure 8.

It is hard for the student not to believe the Central Limit Theorem after you show the fit by plotting a few examples. We fear that this demonstration of the Central Limit Theorem is much more convincing than any proof we have ever given!

A somewhat more difficult program is required to illustrate this theorem empirically by a Galton Board experiment. A demonstration program called **Galton** shows dynamically the wonderful way that an approximate bell shaped curve results (Figure 9).

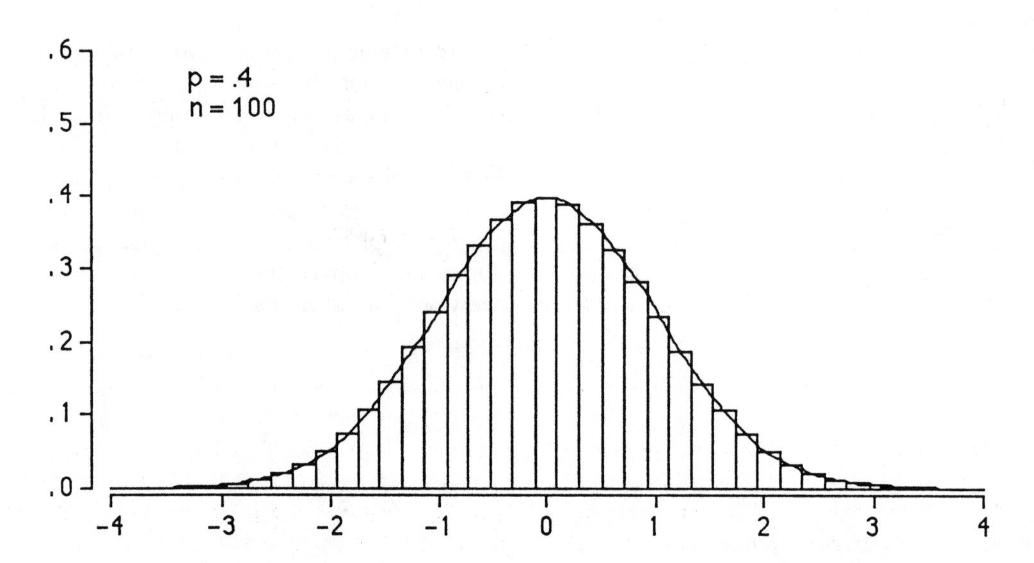

Figure 8. Binomial densities normalized to illustrate the Central Limit Theorem.

Here is the program **Central Limit Theorem** that produced the illustration:

```
! Illustrates the Central Limit Theorem. Plots a histogram of the
! scaled binomial probabilities and compares the graph of the
! approximating normal curve.

library "Lib.prob*"
declare function binomial,normal

set window -4, 4, 0, .5
input prompt "n,p = ": n,p
```

```
let q = 1 - p                                  ! Probability of failure.
let mean = n*p                                 ! Mean of the binomial.
let sd = sqr(n*p*(1-p))                        ! Standard deviation.
for j = 0 to n
    let b = binomial(n,p,j)                    ! Binomial probability.
    let x = (j-mean)/sd                        ! Normalized value.
    box lines x-1/(2*sd),x+1/(2*sd),0,b*sd
                                               ! Plot a bar.
next j
for x = -3 to 3 step .04                       ! Draw normal curve.
    plot x, normal(x,0,1);
next x
end
```

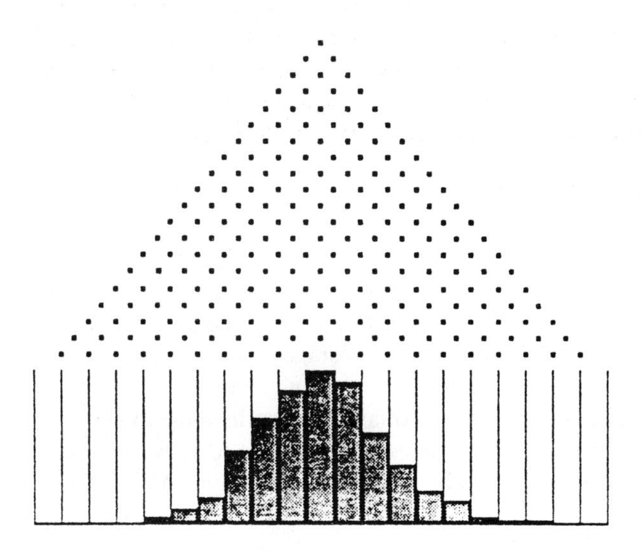

Figure 9. Simulation of the Galton board.

The Central Limit Theorem is one of many examples where the use of computing changes the emphasis on how the subject is taught. For example, the computer really illustrates the fact that the binomial densities converge to the normal density, the so–called local limit theorem, rather than the more conventional version that states that the distributions converge.

The traditional Central Limit Theorem exercises involve learning to transform the given problem mechanically into an equivalent problem in standardized units. Of course, it is important for the student to do a few of these by hand. But by this time it will probably occur to the student to write a program to carry out this conversion. There is little use for tables, and the thoughtful student cannot help but notice that it is easier for him to have the computer carry out a numerical integration process every time he wants to find a normal area than it is to use a table.

Finally, being able to compare the Central Limit Theorem approximations with the exact calculations, with the 1/2 correction, and with the Poisson approximation, the student gets a good idea how good these approximations really are. The fact that exact calculations can be carried out on the binomial density for up to 1000 trials makes one wonder why so much attention is paid to the smallest number of experiments for which the Central Limit Theorem approximation is reasonable.

Having empirical histograms develop dynamically is quite effective in getting the interest of the student. For example, consider a program to illustrate Feller's flying bomb example, in which Feller looked at the distribution of hits of flying bombs over London in the Second World War (one can describe this example less belligerently with raisins falling into cookies). Here it is enlightening to make the empirical count of the number of blocks that get k hits accumulate while the program is running. Running the program for 400 bombs on 100 blocks gave the result shown in Figure 10.

Feller states that this study was made of actual data to see whether it was the case that some blocks got an abnormally high number of hits by chance. The student can see graphically that it is not surprising to have some blocks with more than 8 hits even though the average should be only 4.

Computing can also be used to help the student understand some rather paradoxical aspects of probability. For example, consider Feller's bus paradox. Buses arrive at your corner with exponential time between arrivals with mean 30. You arrive at 12:00

each day; how long will you have to wait for the next bus? This is called a paradox because it would seem that you should arrive about in the middle of the time between buses at your corner and so have to wait about 15 minutes. Yet the answer is 30 minutes.

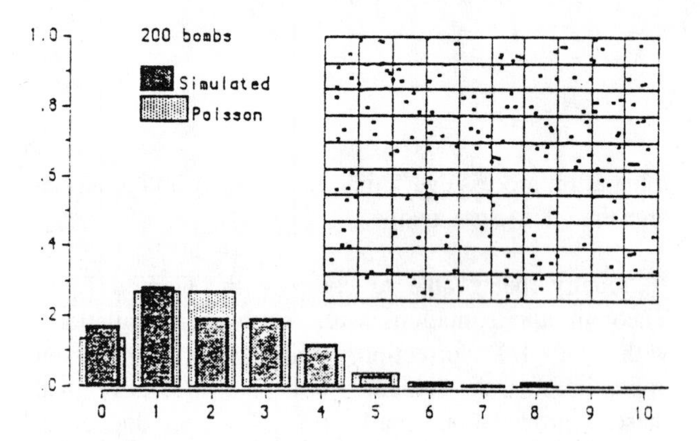

Figure 10. Feller's flying bomb example.

It is instructive to consider a problem like this both for discrete and continuous time and compare the two. As a discrete version, assume that in each minute either 0 or 1 bus arrives, and the probability of an arrival is 1/30. Then the average time between buses is 30 minutes. You arrive at noon; how long will you have to wait on the average? Now the very process of understanding the discrete problem well enough to write the program to simulate this experiment shows the student that the correct answer is 30. His simulations verify this new understanding.

It is instructive to have the student estimate the density for both the time of the next bus and the time since the last bus. This will help explain the seemingly paradoxical situation that buses come to your corner on the average every 30 minutes, but if you were to collect statistics based on your experiences coming every day at 12:00, you would find that the average time between buses is 60 minutes! In Figure 11 we show the results of simulations of waiting times until the next bus compared with the exponential density with mean 30. These empirical densities are quite convincing. A very similar picture is obtained when the student considers the time since the last bus before 12:00, i.e. the time by which you missed the last bus.

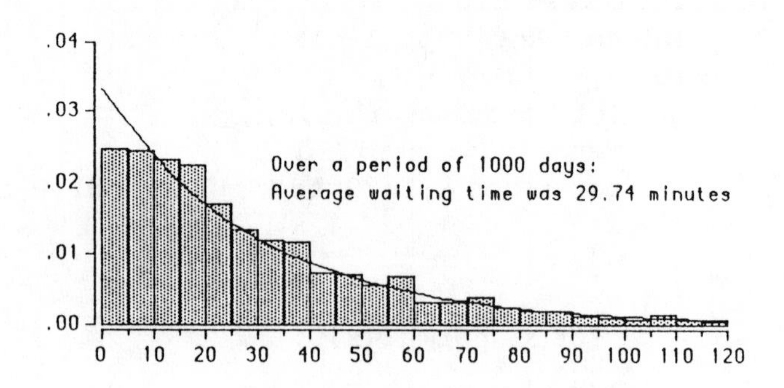

Figure 11. Waiting times for your bus.

A good student will want to know what happens if the time between the buses has some distribution other than the exponential, but still with average time between arrivals of 30 minutes. For example, you might assume that the time between buses is the uniform density on the interval from 0 to 60 or, in the discrete case, the number of heads that turn up when a coin is tossed 60 times.

Queueing theory is a natural subject for using the computer. It is easy to write programs to simulate the simplest kinds of queues to answer some of the standard questions involving the average waiting time, queue size, etc. The qualitative change that takes place as traffic intensity (ratio of average service time to average interarrival time) passes from below 1 to above 1 is very clearly illustrated by simulations, as we see in Figure 12.

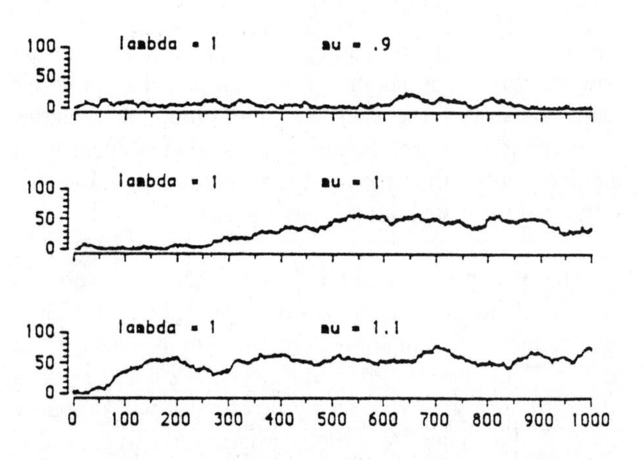

Figure 12. Queue size as traffic intensity changes from less than 1 to greater than 1.

```
sub absorbing chain(tr,Q(),R(),N(),t(),B())

dim I(20,20), X(20,20)                    ! Auxiliary matrices..
dim c(20)                                 ! & vector.

mat I = idn(tr,tr)                        ! tr = number of transient states.
mat X = 1 - Q
mat N = inv(X)                            ! Fundamental matrix
mat c = con(tr)
mat t = N * c                             ! Absorption times.
mat B = N * R                             ! Absorption probabilities

end sub
```

One needs some more advanced computing topics to write good programs that keep track of more detailed statistics of the queues, and at Dartmouth these programs are discussed in our introductory computing science course. This is a nice example of how the two subjects, probability theory and computing science, reinforce either other.

Of course, the computer is a great help in the teaching of Markov chains. First, it is easy to simulate the general finite Markov chain and to illustrate many of the basic results. Second, basic descriptive quantities, such as mean recurrence time, mean time to absorption, and absorption probabilities, can be computed by simple matrix operations from two fundamental matrices easily obtained from the transition matrix. Thus BASIC is particularly good for Markov chains, because the computation of these descriptive quantities requires only simple matrix calculations. A subroutine to compute the fundamental descriptive quantities for an absorbing Markov chain is shown at the top of this page.

Not everyone agrees that students and instructors should write their own programs. John Kemeny points out that expecting the instructor to write programs in order to teach a course is like asking an instructor to write the text for the course. He is in favor of the instructor's doing this if she wants to, but feels that good software should be available for students and instructors to use for demonstration and experimentation without their doing any formal programming. Through Kemeny's efforts, True BASIC provides a probability package for this purpose. The programs are similar to the pretty versions of our programs discussed above. Since the True BASIC language is the same on all machines, such a package is available on each microcomputer on which True BASIC has been implemented. At the moment these are the IBM, Macintosh, Amiga and Atari ST computers. Of course, the programs that we ask the students to write will also run on any of these machines.

Another package of programs for probability and statistics for the IBM computer has been provided by Clark Kimberling. On one disk there are 44 programs that cover descriptive statistics, simulations, counting, discrete and continuous distributions, conditional probability, and random variables. The programs are written in BASICA and are available to the user to modify. The user gets a brief but accurate description of what each program illustrates, and non–trivial examples are often provided.

A second disk provides 33 more advanced programs. A two dimensional plot of a set of random numbers is provided to show that they aren't very random; the user is invited to provide a new random number generator to see if he can do better. Additional programs deal with more specific probability distributions, Bayes' formula, regression, confidence intervals, and nonparametric statistics. The disks contain data files that go with certain of the programs. For example, the student can use data for the number of phone calls on 108 Monday evenings to carry out a chi–square test of fit for the Poisson distribution. The student can also construct her own data files for use with the programs. These disks are available from the University of Evansville Press, 1800 Lincoln Avenue, Evansville, Indiana 47722.

A third interesting package is being developed at the University of Wisconsin by Professor David Griffeath and his colleagues. The project is called GASP (Graphical Aids for Stochastic Processes) and is being implemented on the IBM personal computer; it will be available from Wadsworth and Brooks/Cole. GASP has

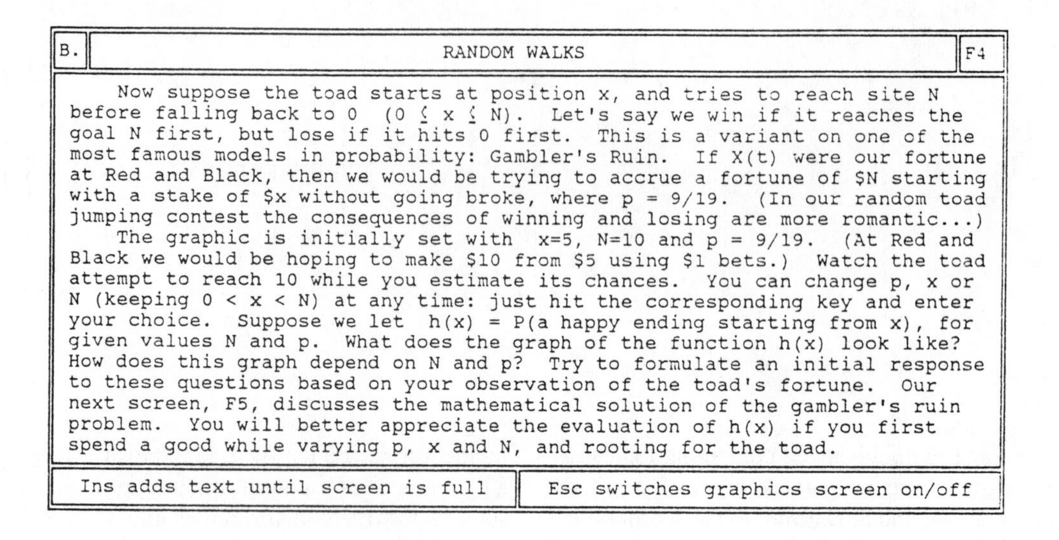

B. RANDOM WALKS F4

Now suppose the toad starts at position x, and tries to reach site N before falling back to 0 (0 ≤ x ≤ N). Let's say we win if it reaches the goal N first, but lose if it hits 0 first. This is a variant on one of the most famous models in probability: Gambler's Ruin. If X(t) were our fortune at Red and Black, then we would be trying to accrue a fortune of $N starting with a stake of $x without going broke, where p = 9/19. (In our random toad jumping contest the consequences of winning and losing are more romantic...)
The graphic is initially set with x=5, N=10 and p = 9/19. (At Red and Black we would be hoping to make $10 from $5 using $1 bets.) Watch the toad attempt to reach 10 while you estimate its chances. You can change p, x or N (keeping 0 < x < N) at any time: just hit the corresponding key and enter your choice. Suppose we let h(x) = P(a happy ending starting from x), for given values N and p. What does the graph of the function h(x) look like? How does this graph depend on N and p? Try to formulate an initial response to these questions based on your observation of the toad's fortune. Our next screen, F5, discusses the mathematical solution of the gambler's ruin problem. You will better appreciate the evaluation of h(x) if you first spend a good while varying p, x and N, and rooting for the toad.

Ins adds text until screen is full | Esc switches graphics screen on/off

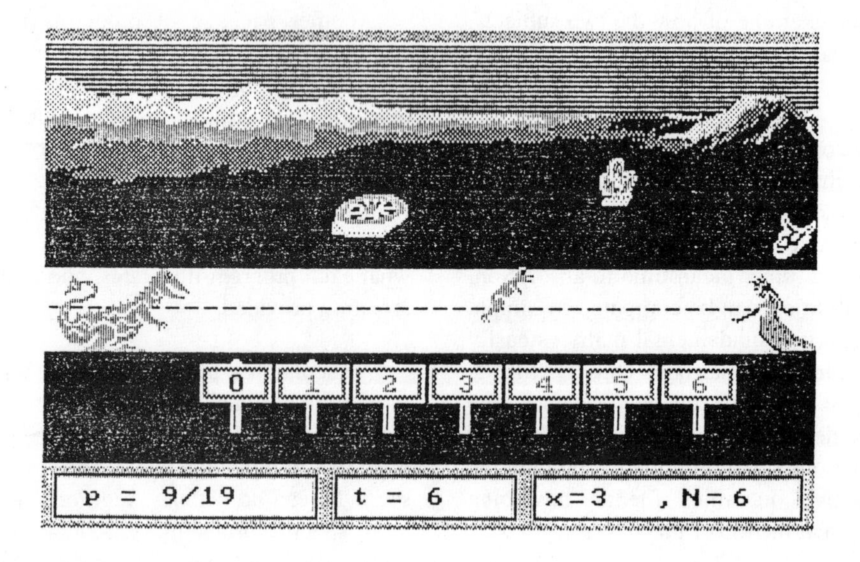

Figure 13. Will the toad reach the princess before the dragon?

several interesting features: First, the authors provide for each example a brief description of the problem, historical remarks, and questions for the student to think about while playing with the graphical output. Second, they use very high quality graphical output, including ideas from computer games, to keep the interest of the student. For example, the discussion of random walk is started with a toad jumping between a dragon and a princess. Static pictures in black and white cannot do justice to the colorful animation in this program, so we leave to the reader's imagination the consequences for the toad (Figure 13) of reaching either edge of the screen.

Having captured the interest of the student in random walks, GASP goes on to give the student a chance to carry out random walks in two and three dimensions and to keep track of $r(t)/t$ where $r(t)$ is the number of states visited by time t. In two dimensions this quantity appears to tend to 0, but in three dimensions it does not, and the student is invited to think about this in terms of trying to decide if a random walker will return to his starting point.

There are many other problems in probability for which animated pictures can stimulate the student's interest. For example, GASP provides the Feller bus problem with actual buses coming by randomly and our friendly toad waiting for them. The user is allowed to vary the arrival times of the toad to see how this affects the waiting times. Feller's bomb problem is illustrated with a realistic picture of London and bursting bombs.

A young boy fishes as two kinds of fish pass by in Poisson streams. The student is invited to learn that the compound process of all the fish is again a Poisson stream, as are the fish that the boy catches (when he catches each fish with a fixed probability). Of course, the user gets to see him catch the fish.

One of the most striking demonstrations shows a Brownian motion path continually crossing the graph of $x = t^{.5}$ but not the curve $x = t^{.6}$. Here the full power of animation is used, as the viewer has the illusion of moving along with the particle, seeing only that last part of the path. It is frightening to think what might happen to probability theory if young and old get hooked on these clever animations.

Returning to our work at Dartmouth, computer simulations are used to illustrate problems that are current areas of research in probability. Examples include the Ising model, growth models, and a great variety of interacting particle models. One that always creates an interest among the students is the *stepping stone* model from genetics. In this model, we have an array of squares, each initially one of, say, three different colors. Each second a square is chosen at random. This square then chooses one of its eight neighbors at random and colors it to agree with its color. Figure 14 shows a random initial configuration of three colors and then the result after the process has run for some time.

The student sees that territories are established, and it becomes a battle to see which color survives. At any time, the probability that a particular color will win out is equal to the proportion of the squares of this color. This is because the number of squares that are a given color is a fair game (martingale). Thus this example can also be used to show the power of system theorems from the gambling interpretation. The simulation itself could be made into an exciting gambling game, since there is always much interest in which color will win.

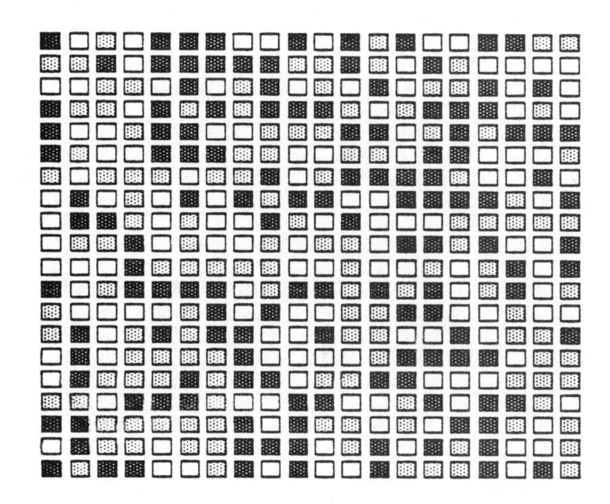

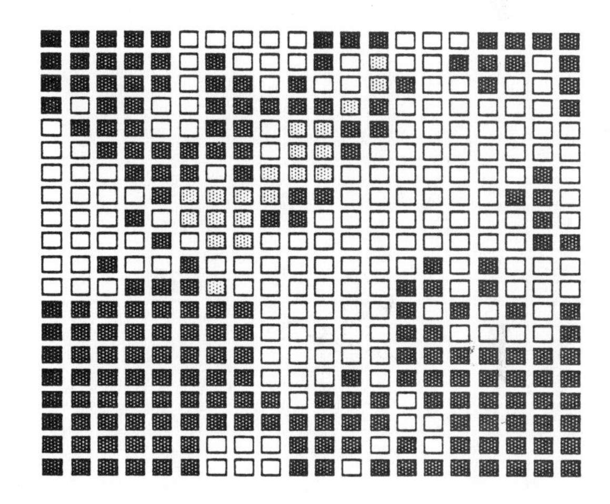

Figure 14. Simulation of the stepping stone model from genetics.

We have tried to show how we use computing in a probability course. As we have emphasized, our approach is to integrate the computing into the course and to expect the instructor and the student to do simple programming. Of course, this requires that the student have easy access to a computer. It also requires that the classroom be equipped with high quality computer display devices — either monitors or a projector. The fact that serious probabilists like David Griffeath and John Kemeny are willing to work on the software to make probability come alive in your classroom should be encouragement enough: Give it a try; you'll like it!

REFERENCES

1. Feller, William. **Introduction to Probability and its Applications**, Vol. 1, 3rd ed. New York: Wiley, 1968.

2. Snell, J. Laurie. **Introduction to Probability with Computing**. New York: Random House–Birkhäuser, to appear.

A Logo–Based Course in Problem Solving

Charles A. Jones

The goal of this essay is to provide ideas about how and why Logo is used in teaching college mathematics courses. The first two sections consist of generalities about Logo and about its use in college mathematics specifically. The third section, which forms the bulk of this essay, is a more detailed look at the "Problem Solving and Computing" course at Grinnell College, which makes extensive use of Logo. I developed this course with support from the Alfred P. Sloan Foundation's "New Liberal Arts" program. The course is, roughly speaking, an attempt to merge problem solving ideas, such as those of George Polya [6] and Alan Schoenfeld [7], with Logo ideas, such as those of Seymour Papert [5] and Harold Abelson [1].

Logo and the educational philosophy behind it.

Logo is a high level programming language that features turtle graphics. Because of the graphics capabilities, one can quickly learn to develop interesting programs. Other advantages of using Logo as a teaching language include a built–in editor and the natural way in which procedures are used to build up longer programs. In addition, Logo is a powerful language, with lists as first class data objects, that is well suited for writing sophisticated programs.

One of the major goals in using Logo is to establish an active learning environment in which each student can control much of the learning process — the student programs the computer rather than being programmed by the instructor or a computer. A related goal is to correct the notion many students hold that mathematics is best learned in a dissociative manner. In fact, one hopes to actively alter the current situation in which many of our students do indeed dissociate themselves from the learning of mathematics. My favorite passage from *Mindstorms* [5] is a quote attributed to a fifth grader describing his learning technique for mathematics: "You learn stuff like that by making your mind a blank and saying it over and over until you learn it."

Those of us who teach college mathematics have to cope with the same syndrome of students learning course material by memorizing and mimicking but neither trusting nor using their own creative thinking abilities. An educational goal of Logo is the development of each student's confidence and ability to use creative thinking in the realm of mathematics. Papert emphasizes strongly the benefits of "Piagetian learning, which is to [him] learning without curriculum." Briefly, Piagetian learning means exploring and creating (in a supervised, but not goal–oriented, environment) in order to acquire experiences. These experiences are then used in each individual's development of intellectual structures and theories.

Logo at the college level.

Logo is being used at several colleges and universities as a language for programming courses and as a language to teach mathematical concepts. One important reason for this is the wealth of interesting problems enabled by its graphics capabilities. The process of solving these problems, by building programs to produce elaborate displays from smaller procedures which produce simple pictures, is important both from a problem–solving/algorithm–development point of view and for the discovery of geometrical ideas.

An even more important theme is that of experimenting and correcting mistakes. Students learn a worthwhile and novel lesson: one can experiment in mathematics and be creative when solving problems (as opposed to merely regurgitating facts and techniques already practiced). It is equally important for students to discover that, if an attempt at solving a problem is not successful, they can *think* about what went wrong and attempt a correction.

Courses using Logo are also achieving some success in overcoming math or computer phobias. My own experience indicates that this has happened for a few students. For many others it would be more accurate to say that their attitudes towards and perceptions of mathematics have changed. The fundamental reason for this change is the realization that their thinking processes can be brought to bear on mathematical problems. The most valuable lesson of such a course may be this (partial) triumph over the attitude that mathematical subjects are best learned in a dissociated manner.

The Problem Solving and Computing course at Grinnell.

Background and goals. The student for whom this course is designed is a nonscience–oriented, bright undergraduate who writes well and thinks well in nonscience disciplines. Problem Solving and Computing is our department's "Math for poets" offering.

There are two major topics for this course: mathematical problem solving and algorithm development and programming. These two topics are definitely complementary. The second can be regarded as a subtopic of the first, but each is addressed separately, with the connections between them continually being examined. In fact, the problem solving heuristics are more easily and more consistently applicable to programming than to the broader, more amorphous topic of problem solving.

The textbooks for the course are *Thinking Mathematically* [4] and *Turtle Geometry* [2]. *Thinking Mathematically* works very well as a problem solving textbook, although some students find its tone condescending. *Turtle Geometry* is an excellent but very difficult text; while the ideas are fascinating, the book overwhelms too many in the intended audience for the course.

The goals of the course, in decreasing order of importance, are:

To improve each student's ability to deal with problems in any setting, especially an unfamiliar one.

To serve as an introduction to programming and algorithm development.

To give the students a taste of *doing* mathematical activity.

To help the students acquire a new perspective on the nature of mathematics.

To have the students learn some mathematical concepts from geometry and elementary number theory.

Course Specifics. The course meets three hours per week in a regular classroom equipped with an Apple II computer connected to large display monitors. There is a one hour computing lab each week in a room with about 10 Apple computers.

The classroom time is divided among doing problem solving, doing algorithm development, discussing problem solving readings, lecturing on mathematical and computer science topics, and answering questions. The "doing" portions usually consist of a program or problem being posed by the instructor and the class working on it en masse. The discussions of *Thinking Mathematically* center on both the students' opinions of the validity of the heuristics presented and the students' experience in applying these heuristics in their own work. It is not necessary to lecture on the material in this book, as the students find it easy to read. The lectures cover such topics as the assignment statement, parameter passage, recursion, greatest common divisors, and prime numbers.

In theory, the students have discovered patterns occurring in their problem sets, programming assignments, and labs. The lectures then organize these patterns into conceptual frameworks. As one might expect, this aspect is only moderately successful, and of course the amount of success varies greatly from one student to another.

The computing lab segment of the course consists of the students being assigned fairly simple programs to write with the instructor roving about and giving individual attention. This aspect of the course is very successful. One reason is that most of the Logo syntax is introduced here in a "hands on" environment; the course does not use a Logo textbook. A more important reason is that the instructor is serving as a "coach" rather than as a "demonstrator" (see [7]). The students are actively engaged in the problem solving process, and the instructor is helping them with *their* methodology. The labs also accomplish such tasks as developing key procedures for the students' assignments and developing standard constructs not in Logo (e.g., the while loop).

In addition, the labs are an ungraded environment in which students can explore with Logo while they have access to the instructor. They are encouraged in the lab sessions to go off on tangents or change the suggested problem. It seems impossible to have "Piagetian learning" in a college level course for which grades must be assigned, but it is worthwhile to strive for some of that type of learning.

Assignments. This is the most important aspect of the course, as the whole philosophy of the course is based on "active learning". The students are assigned (approximately) weekly programs, weekly problem sets, and an end–of–the–term paper about problem solving. The programming assignments listed below illustrate some of the advantages of Logo in a problem solving course. Many of the ideas for these assignments come from [1] and [2]; the babysitter problem is due to Henry Walker.

Task: Drawing the Olympic Logo.
Purpose: Top–down design, use of procedures.

Task: Drawing nested rectangles.
Purpose: Procedures with inputs (parameters).

Task: Drawing polygonal spirals.
Purpose: Recursion, use of if–then, modification of the known polygon procedure.

Task: Tree drawing.
Purpose: Recursion.

Task: Drawing a random walk.
Purpose: Recursion and the use of a random number generator.

Task: Counting characters.
Purpose: Recursion and functions.

Task: Deciding if a given input is a palindrome.
Purpose: Recursion and functions.

Task: Computing a babysitter's fee, given the starting and ending times, where the hourly rate varies by time of day.
Purpose: Functions and composition of functions.

Task: Simulating animals chasing one another.
Purpose: Simulation and recursion.

Task: Drawing Hilbert Curves.
Purpose: Recursion, strong tie to induction, modification of a different Hilbert curve program.

Task: Project.
Purpose: Larger program, top–down design.

Logo is sometimes referred to as "Lisp with pictures," and the "pictures" are a great advantage. Procedures are introduced at the start of the course and recursion by the third week. This would not be reasonable without the graphics. Recursion (my favorite programming topic) is a natural tool to use for drawing the pictures in some of the early assignments. This tool can then be exploited for list processing problems and more advanced recursive designs. (When the students are first using recursion I avoid using the terminology "recursion" or "recursive procedure," so they will not be inclined to regard the technique as anything other than the natural way to proceed.) Recursion is a very successful topic; all of the students can use it at least adequately by the end of the term, and the better students have a powerful technique at their disposal.

Another advantage of Logo is that programs are very easy to modify by adding procedures but making no changes to the existing code. This feature allows open–ended assignments for which everyone does the basic program, and the better or more motivated students can add on features.

The problem sets and the problem solving heuristics are based mainly on [4], [6] and [7]. Near the end of the semester, the students each write a paper about problem solving. In addition, they turn in write–ups with each assignment and after some of the labs. Each write–up includes such items as what was hard about the problem, where they got stuck and how they become unstuck, and what design decisions they made. The rationale for these writings is twofold. First, they allow the students (especially the weaker problem solvers) to show what they've learned *about* problem solving and computing techniques. Second, this writing causes the students to practice a good problem solving technique, that of "looking back" or "reflecting".

Conclusion.

While most of this essay has been very positive about Logo, college courses using Logo face some difficulties and compromises. For example, in the Grinnell course there is not much "Piagetian learning," and there is not as much mathematical content as one usually expects in a college level course. However, from interviews with some of the best and/or most enthusiastic students approximately one year after the course, it seems that all the goals of the course, except for learning mathematical concepts, were well met. Even though I am not happy about the lack of retention of mathematical concepts, I believe meeting the other goals makes this course a worthwhile liberal arts offering. In some schools greater emphasis is placed on learning mathematical material and concepts. For example, Logo is being used as an integral part of the mathematics instruction in teacher training courses at the University of Montana [3].

Logo was designed to be the pioneer in an evolving family of computer languages. While Logo itself is an excellent tool, we can look forward to more powerful tools with the same underlying philosophy in the future.

REFERENCES

1. Abelson, H. **Logo for the Apple II**. New York: Byte/McGrawHill, 1982.

2. Abelson, H., and A. diSessa. **Turtle Geometry**. Cambridge: MIT Press, 1981.

3. Billstein, R., S. Liebeskind, and J. Lott. **A Problem Solving Approach to Mathematics for Elementary School Teachers**. Menlo Park, CA: Benjamin/Cummings, 1981.

4. Mason, J., L. Burton, and K. Stacey. **Thinking Mathematically**. Reading, MA: Addison–Wesley, 1985.

5. Papert, S. **Mindstorms: Children, Computers, and Powerful Ideas**. New York: Basic Books, 1980.

6. Polya, G. **How to Solve It**, 2nd edition. Princeton, NJ: Princeton University Press, 1957.

7. Schoenfeld, A. **Problem Solving in the Mathematics Curriculum: A Report, Recommendations, and an Annotated Bibliography**, MAA Notes #1. Washington, DC: Mathematical Association of America, 1983.

Geometry and Computers

James R. King

A computer is a flexible and powerful drawing tool. It can also be a tool for thinking. Therefore, computers have great potential for teaching geometry. However, when one sets out to teach a geometry course using computers, there is no well–trod path to follow. There are many choices to make, both about the mathematics to include and about how the computers will be used. This paper is a discussion of some topics in geometry that lend themselves to computer exploration. It is based on my experience teaching a course in which the computer is the central tool, with students doing their own programming in Logo.

My view is that, while there is no separate subject of computer geometry, the use of a computer can illuminate geometric ideas that are not as tractable by hand. We already see this phenomenon in the experience that some geometric ideas seem clearer using synthetic methods, while some are easier to think about using analytic methods. It is also true that using both approaches may show more than one face of the same idea. The computer gives a somewhat different view of geometry, which sometimes may not add much, but other times may lead to a very different way of thinking about a geometric situation.

Sometimes the contribution of the computer is not so much a new approach as the facility to generate rapidly a host of complex examples. However, this facility can give a subject a whole new flavor. A rich stock of examples leads to better understanding. When one can generate families of lines as easily as one could formerly generate individual triangles, one can take a more experimental approach to subjects that were formerly accessible only to formal argument.

The method of turtle geometry.

The drawing tools of the ancient Greeks were compass and straightedge, and the use of these tools shaped the thought in their geometry. With a computer one can draw by simply plotting the (x, y) coordinates of analytic geometry, but while this can be a valuable approach on occasion, it loses the feel of drawing that is present in classical Euclidean constructions, and it may provide little geometric insight. There is, however, a flexible and truly geometric drawing tool for a computer; it is called, curiously, a turtle. The geometry based on this tool is called *turtle geometry*.

An element of plane turtle geometry is not just a point, it is a *turtle state*, which is a pair consisting of a position (a point in the plane) and a heading (which can be represented by an angle or a unit vector). The turtle drawing tool appears on the computer screen as an arrow located at the position and pointed in the direction of the heading. The drawing commands are a forward command, which moves the turtle in the direction of the heading and draws a line from the old position to the new one, and a turn command, which changes the heading.

Figures are drawn by a succession of these commands, using the logical organization afforded by the presence of repetition, logical branching, and perhaps recursion in the computer language. Originally, the turtle was a robot that crawled along the floor, drawing with a pen; it looked something like a turtle, hence the name turtle geometry. The turtle in a computer representation can also (metaphorically) raise and lower its pen, depending on whether one wants it to draw when it moves or not.

To use the computer turtle, one gives it commands. The command `FORWARD 100` causes the turtle to move its position 100 units in the direction of the heading, while leaving the direction unchanged. The command `LEFT 90` causes the heading to be rotated counterclockwise 90 degrees, leaving the position unchanged. Thus a sequence of four commands `FORWARD 100 LEFT 90` draws a square, while a sequence of three commands `FORWARD 100 LEFT 120` draws an equilateral triangle. Also, in Logo and other suitable languages, it is possible to write a procedure, e.g., `SQUARE` or `TRIANGLE`, that incorporates the drawing commands for a figure and can be invoked as a new command.

Like most powerful ideas, this is a fascinating mixture of the simple and the profound. Turtle geometry is simple in the sense that ruler and compass constructions are simple; it is accessible to our intuition and builds on our geometric experience. Learning to draw with the turtle is easy enough that young children can learn to draw complex figures. Giving instructions for drawing a figure is very much akin to giving directions to someone driving a car ("You can't miss it; just go straight ahead a quarter mile, then turn left ..."). The profound part is that this geometry is an intrinsic, coordinate–free geometry that unites classical polygons with ideas from modern differential geometry. It turns out to be well–suited for all sorts of geometric investigations, as we shall see.

An elementary consequence of the intrinsic nature of turtle geometry is that one can draw the pinwheel in Figure 1 without computing any (x, y) coordinates. Just repeat the commands `SQUARE RIGHT 36` ten times. Other examples of the value of an intrinsic approach to geometry will appear later.

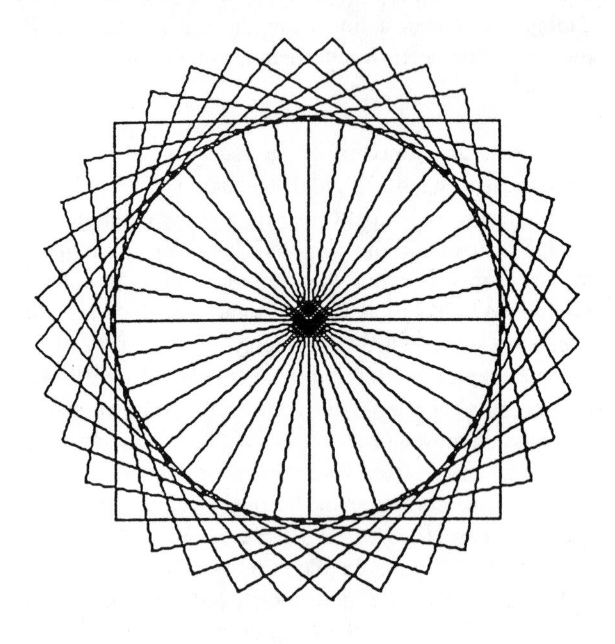

Figure 1. Pinwheel:
`REPEAT 10 [SQUARE RIGHT 36].`

Turtle geometry is often associated with the Logo programming language, for it was introduced and advocated by the group at MIT around Seymour Papert that also developed Logo. The merits of Logo as a language for learning mathematics are presented eloquently by Papert in *Mindstorms* [11]; however, the turtle approach to geometry can be implemented in virtually any computer language, although with some limitations. This is explained in the fundamental reference and text by Abelson and diSessa [1].

Foundations of turtle geometry.

Before proceeding with turtle geometry on computers, I digress to say a few words about its mathematical foundations. Mathematically, an element of the turtle geometry of the plane is a turtle state, a pair consisting of a point in the plane (the position) and a direction (the heading), which we represent by a unit vector. Since the set of unit vectors in the plane forms a circle, the space of all turtle states is a three–dimensional space, the cartesian product of a plane and a circle. There are two natural one–parameter groups of transformations on this space, the forward transformations F_t and the turns L_u.

Let F_t be the transformation that carries any state (P,V) to the state $(P + tV,V)$. Here P is a point, V is a unit vector, t is a real scalar, and addition is vector addition. Let R_u be counterclockwise rotation by angle u about the origin. Then let L_u be the transformation that maps (P,V) to $(P,R_u(V))$. These transformations correspond to the Logo commands `FORWARD` and `LEFT`. A general turtle geometry transformation is the composition of these simple transformations. It is not difficult to show that a general turtle transformation T is given by $T(P,V) = (P + MV,RV)$, where R is a rotation R_u, and M is a linear transformation of the form $M(x,y) = (ax - by, bx + ay)$, i.e., M is a scalar multiple of a rotation; if the scalar is not 0, this is a similarity transformation. Such transformations can also be represented by complex numbers. Thus turtle geometry can be studied by means of its group of transformations, in the spirit of Felix Klein's approach to geometry. This group of turtle transformations is a 3–dimensional group.

It is interesting to note that the proper Euclidean group on the plane also defines a 3–dimensional group on the state space but it is a different group. The transformation $S(P) = AP + B$, where A is a rotation matrix and B is a position vector, induces the transformation that maps (P,V) to $(AP + B, AV)$. These two groups commute with each other and have only the identity in common. The action of each group is transitive and free on the state space. In other words, if two states are given, there is exactly one transformation in each group that carries the first to the second. This means that the transformation group can be identified with the state space, provided a reference, or "home," state is chosen. The best reference for the foundations of turtle geometry and turtle transformations is [1], although the turtle transformation group itself is not discussed as such there.

Symmetry and total curvature.

The topic most accessible to the beginning turtle geometer is rotational symmetry. It is easy to draw a rotationally symmetric figure; just repeat any sequence of turtle transformations over and over until the turtle returns to its original state. (This is a slight over–simplification; the correct statement is given below.)

As an example, we consider the simplest turtle geometry procedure, called `POLY` in the Logo literature. `POLY` commands the turtle to go `FORWARD` a distance of S units and turn `LEFT` A degrees, where S and A are inputs, and then to repeat indefinitely. If A is $360/N$, the procedure draws a regular N–gon. For A a rational multiple of 360, say $A = 360\,p/q$, with p/q reduced to lowest terms and $q > 1$, `POLY` draws a star polygon with q vertices (see Figure 2).

The conventional approach to rotational symmetry is to use the center of symmetry Q. The rotational symmetry of the figure is described by the group of rotations with center Q that leave the figure invariant. To draw a regular polygon or star polygon, one would mark off q equally spaced vertices on a circle centered at Q. By connecting consecutive points on the circle, one draws a regular polygon; by connecting every p–th point, one draws star polygons. The turtle geometry approach is to observe that in this construction the

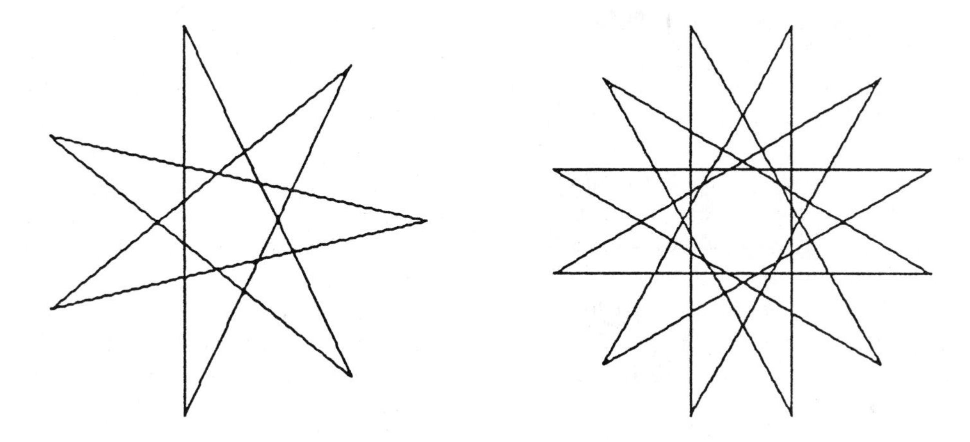

Figure 2. Star polygons drawn by POLY.

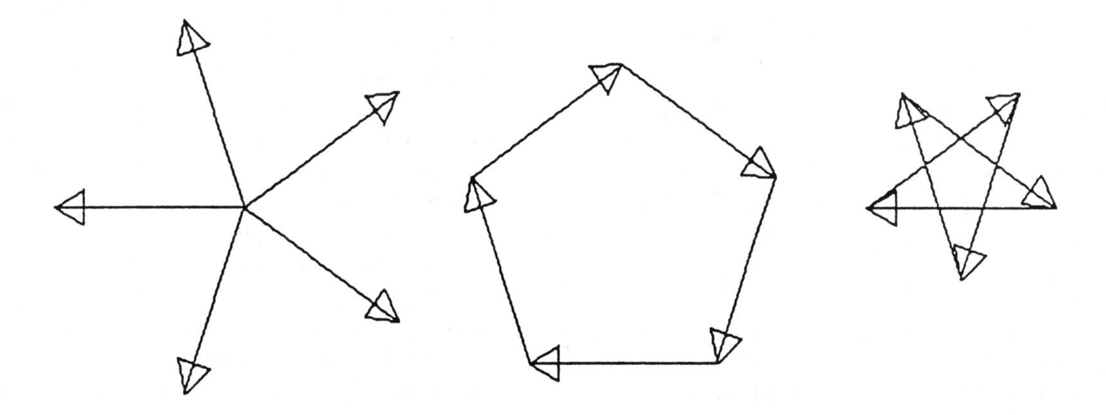

Figure 3. The center of symmetry for a polygon and star polygon.

direction vectors from one vertex to the next also are equally spaced around the circle of unit vectors, so one draws the figure by turning the turtle in these directions by a sequence of equal turns. (See Figure 3.) One feature lost with this approach is that the center of rotational symmetry of the figure must be computed independently, but a very important gain is that it is possible to create very complex rotationally symmetric figures with ease.

The tool that makes this possible is the *Total Turning Theorem* and the concept of *total turn* of a turtle path. The total turn of a turtle path is simply the sum of the degrees of all the left turns in the path (where a right turn by A degrees is viewed as a left turn by $-A$ degrees). The Total Turning Theorem states that, for a closed turtle geometry path, the total turn is an integer multiple of 360. A companion theorem says that for any simple closed path, i.e., a path without self–intersections, the total turn is ± 360; the integer multiplier is ± 1.

This can be used to analyze figures. For example, in Figure 4 the 7–pointed star traced in a counterclockwise direction has a total turn of 360. If the interior angle at one of the outer vertices is 30, then the turn at this vertex is $180 - 30 = 150$. But the figure is made up of 7 congruent parts, each consisting of two segments and two turns. The total turn of each part must be 360/7, as the sum is 360. Thus the turn at the inner, reentrant, vertices must be $(360/7) - 150$.

Figure 4. A complex star.

The flower in Figure 5 has five petals, each of which is tangent to its neighbors. Since they are tangent, the interior vertex angle of each petal is $360/5 = 72$. On the other hand, each petal has a 2–fold rotational symmetry, i.e., rotation by 180 degrees, so each of the two congruent pieces consisting of an arc and a turn must have a total turn of 180. Since the turn is $180 - 72 = 108$, the arcs in the petals must have central angle 72 degrees.

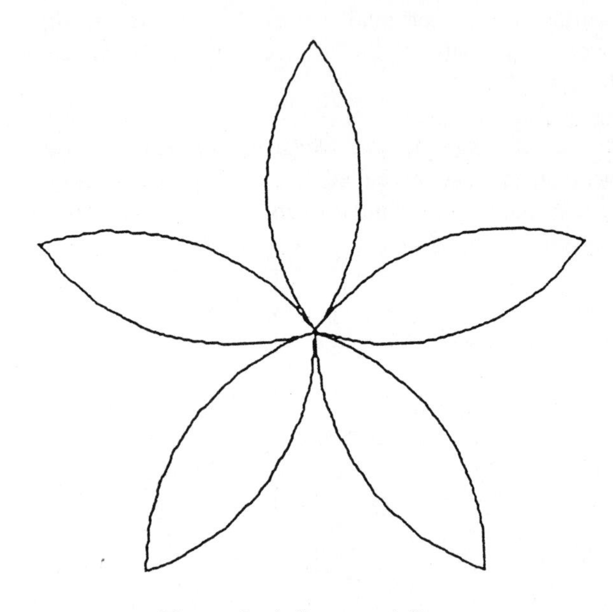

Figure 5. A five–petal flower.

The careful reader may object at this point that arcs are not polygons, and this figure is not covered by the Total Turning Theorem. There are two ways to answer this objection. From the computer point of view, arcs are drawn as parts on regular n–gons for large n; for example, this "Logo arc" could be (and was) drawn as 72 sides of a 360–gon. However, what makes this more interesting is the second point of view, that the theorem can be extended to arcs and in fact to piecewise smooth curves in general, where the total turning is replaced by the *total curvature*, the integral of the curvature of the smooth curves with respect to arc length, plus the sum of the turns where the curve is not smooth.

It is a theorem of differential geometry that the total curvature of a closed curve is a multiple of 360 degrees (although it is usually expressed in radians). For an arc of a genuine circle, the total curvature is precisely the central angle subtended by the arc, and it is approximately equal to the total turning of a polygon used to approximate the arc. We note also that such piecewise smooth curves define curves in the state space: If $X(s)$ is a curve parametrized by arc length, the curve $(X(s), X'(s))$ defines a curve in state space; this is a standard technique in differential geometry. By adding turn commands at the corners of the piecewise smooth curve, one can define a continuous curve in the state space and, by projection, a continuous map to the unit circle. Then the Total Turning Theorem becomes a statement about the topological properties of such maps, which are classified by an integer called the *degree*, which in this case is the total curvature divided by 360 degrees.

However, the use of the total turn in this way is not confined to the analysis of existing figures. Suppose that a figure is drawn by repeating a basic sequence of turtle commands indefinitely. Suppose that the total turn of the basic figure is $360\,p/q$ degrees, $q > 1$. Then q repetitions of the basic sequence of commands draws a closed turtle figure. In other words, if the total turn after repeating the basic figure a certain number of times is an integer multiple of 360, then the figure has closed up. Thus for such looping figures, the converse of the Total Turning Theorem is true.

This converse means that figures can be created by taking the total turn into account. Thus, the analysis of the two figures above is sufficient to create the figures as well. But other figures can result from similar analysis. For example, the star in Figure 4 has equal sides defining each vertex. If one repeats instead the sequence of commands `FORWARD S LEFT 150 FORWARD T LEFT 360/7 - 150`, then one gets a star which is no longer symmetric with respect to reflection in a line. (See Figure 6.)

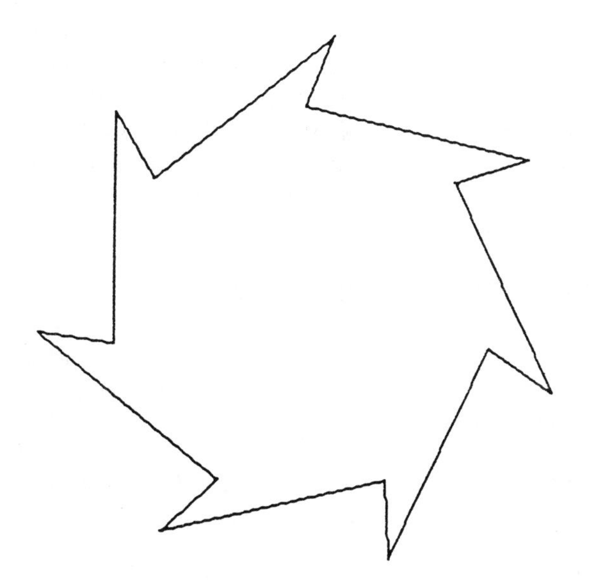

Figure 6. A more complex star.

One can also create sometimes beautiful figures by simply repeating some random jotting until it closes up. In Figure 7 the doodle on the left is repeated to form the symmetric figure on the right.

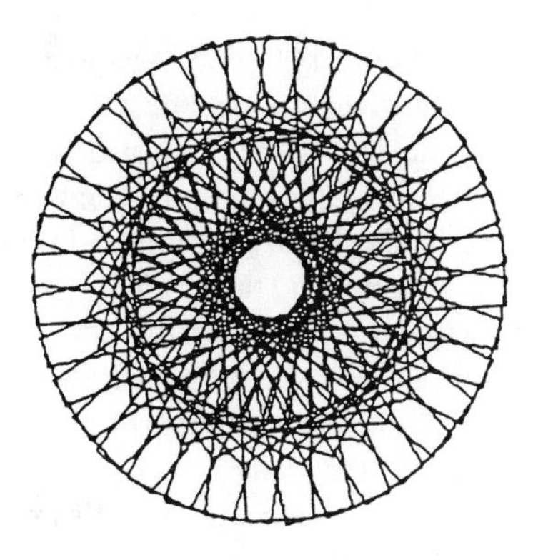

Figure 7. A repeated doodle pattern.

In a course one can explore a great deal of interesting geometry with this approach. For example, one can:

- create and analyze figures made up of lines and arcs and describe their rotational symmetry;
- write an algebra procedure to reduce fractions to lowest terms, using the Euclidean algorithm for the greatest common divisor, and use this to determine the number of repetitions in looping programs;
- prove all the standard theorems of high–school geometry about angles and secants and chords of circles by taking turtle trips around the figures;
- compute the centers of figures;
- inscribe polygons by connecting corresponding points of looping figures;
- use recursion to create more complex forms of repetition (e.g., the procedure INSPI, which changes the angle of the turn in a periodic way [1]).

According to one's preferences, all this work can be done as informal but challenging geometry using computer programming alone, or it can be accompanied by a traditional treatment of symmetry with the degree of rigor that seems appropriate. The most complete reference for the material of this section, and the only one with proofs, is [1], but almost every Logo book discusses these ideas on some level.

Turtle geometry and spirals.

Spirals are another very satisfying topic to explore with turtle geometry, since they are easy to generate and create spectacular designs. They are created very much in the style of `POLY`. One common form of spiral is called `POLYSPI` in the Logo literature. In addition to inputs of the side length S and the angle A, the procedure has an increment input I. A recursive definition of the procedure is to go `FORWARD S`, turn `LEFT A`, then run the procedure again with the side length changed to $S + I$ and the other inputs the same. Thus the figure is composed of turns by a constant angle followed by sides whose lengths increase at a constant rate. (See Figure 8.)

Figure 8. A `POLYSPI` pattern.

One interesting effect is created by choosing the turn angle to be a rational multiple of 360 with small denominator. Then the spiral appears as a growing polygon. On the other hand, if the angle is chosen slightly different from such an angle, a precession phenomenon is obtained. One can experiment to obtain optical illusions and moiré patterns.

The `POLYSPI` spiral does not have any symmetry transformations that are similarity transformations of the plane. A spiral that does is the equiangular or logarithmic spiral. This is obtained in turtle geometry by making the new side length a constant multiple of the preceding one. This spiral appears in a number of biological models: By building a chain of similar figures (e.g., triangles or quadrilaterals), one obtains rather natural–looking shells or horns, and the procedure for creating them mimics the process that occurs in nature. (See Figure 9.)

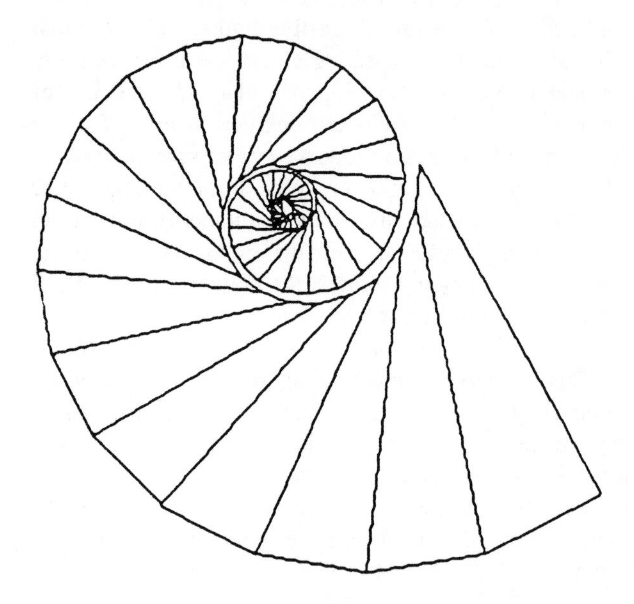

Figure 9. A logarithmic spiral of quadrilaterals.

One strength of turtle geometry in this kind of investigation of complicated figures is that is quite easy to study how the figures change if parameters such as angle or ratio are changed. This leads to thinking about deformation, stability, and other dynamic aspects of geometry [1], [15].

Turtle geometry and fractals.

Fractal geometry is an excellent topic for a computer geometry course. It is exciting because it is surprising for students and because it is an active area of contemporary mathematical research. Even better, a computer makes this sophisticated mathematics accessible for experimentation by undergraduates.

Since turtles draw curves, the part of fractal theory that works best with turtle geometry is the theory of fractal curves. These are curves topologically, but metrically their dimension is greater than one. An extreme example is the Peano curve, which fills a square region, but examples that fill less space can have quite beautiful geometry. For example, the Koch snowflake is obtained as the limit of a sequence of curves. The first curve is a triangle, and the second is obtained by replacing the straight line segments in the triangle by congruent lumps consisting of four segments. At each successive stage, the next curve is obtained by replacing every segment in the previous curve with lumps similar to the first ones. The limiting curve is a curve of Hausdorff dimension great than 1. On the computer one must stop before infinity, but after a number of stages a curve of great beauty and complexity is obtained. (See Figure 10.)

Figure 10. A stage of the Koch snowflake.

Using this pattern of taking a figure and a model lump and recursively replacing every segment in the figure by a lump similar to the model, it is possible to obtain an extremely varied collection of curves of mind–boggling complexity. The process of actually writing a computer program to draw such a curve requires the student to really understand the recursive construction, and the feedback from trying out the program makes it possible to correct possible misconceptions. Turtle geometry is almost essential to the writing of this sort of program, for it is simple to describe the lump with turtle commands and then draw it at any size and at any location in the plane, whereas the use of Cartesian coordinates leads to very difficult and opaque computations.

Examples of fractal curves can be found in [1] and [15], but anyone interested in fractals must refer to the fundamental book [10] for the explanation of the theory of fractals, for the magnificent collection of pictures (which can serve as a source of countless experiments), and for its bibliography. The examples of curves that are limits of polygonal paths are the most accessible to turtle geometry, but using a computer with fast graphics commands to draw arcs and fill shapes, one can approach some of the other examples as well.

Tilings.

The study of tilings of the plane is much more accessible to students using the traditional tools of geometry than is the study of fractals, but it can also benefit from the introduction of computer drawing, because reproducing a tile over and over is so labor–intensive, and because the patterns in the logical structure of the program echo the mathematical pattern of the tiling. With a computer one can complement the proof of the theorem about which regular n–gons tile a neighborhood of a point in the plane by writing a program which actually draws the successful and unsuccessful cases on the screen. Understanding how to determine which n–gons to try in the program and when all the possibilities are exhausted is very close to understanding the rigorous proof.

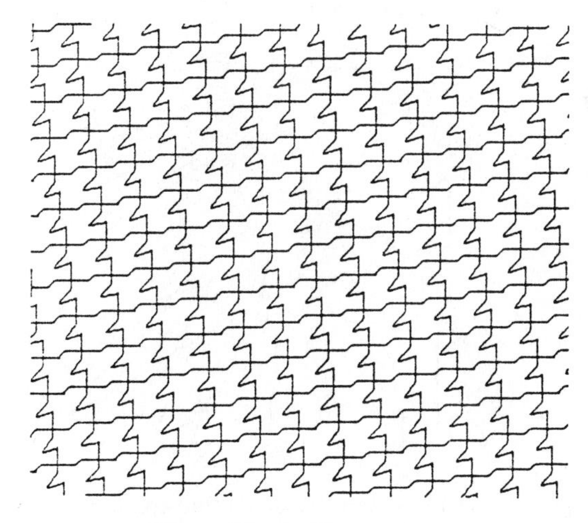

Figure 11. A tiling pattern.

One can write programs that exhibit the standard tilings of the plane, then modify the tiles in some artistic way to get patterns inspired by the drawings of Maurits

Escher or by the designs of the Moors. (See Figure 11.) There are examples of this in [1] and [15], but richer sources of ideas are [6] and [14]. There are also some books, such as [2], intended for younger students, with exercises for pencil and paper that can be converted to computer programs.

Line conics and string figures.

There are numerous curves of classical geometry and modern geometry that are defined as the envelope, or outline, of a family of lines. One example is the construction of conics as line conics. The simplest such constructions are done at the kindergarten level with string. A child creates a string figure on some pattern, say the two sides of an angle drawn on cardboard, by punching holes regularly along the pattern and then connecting the n–th hole of the first side to the n–th hole of the second side by sewing with string or yarn. This makes a pretty pattern. The outline, or envelope, of the strings is a parabola. (This is a special case of a theorem of projective geometry that states that if the points of two lines are related by a projective transformation, the lines connecting corresponding points form a line conic.)

It is easy to simulate this child's play with turtle geometry. One moves the turtle along each line and

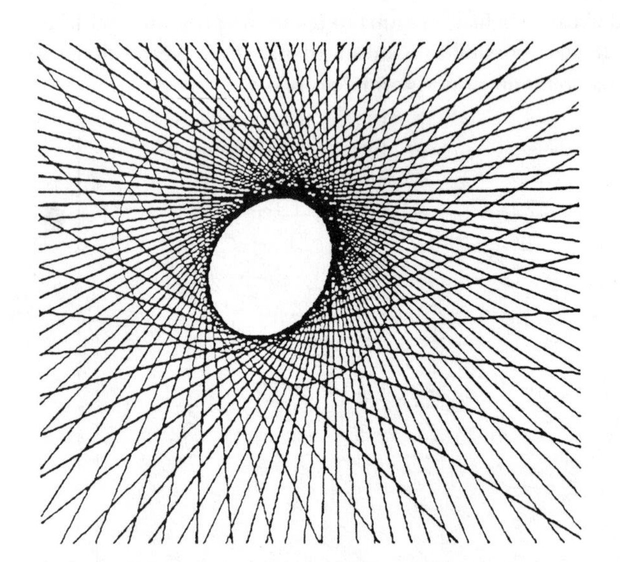

Figure 12. An ellipse as a line conic.

saves a list of points for each. Then one draws a line through each pair of corresponding points. Now with essentially the same program, one can draw other string figures that produce ellipses and hyperbolas (see Figures 12 and 13). We run the turtle around two circles, marking points and connecting them as before. Depending on the relative positions of the circles and the starting points, conics of varying shapes are obtained.

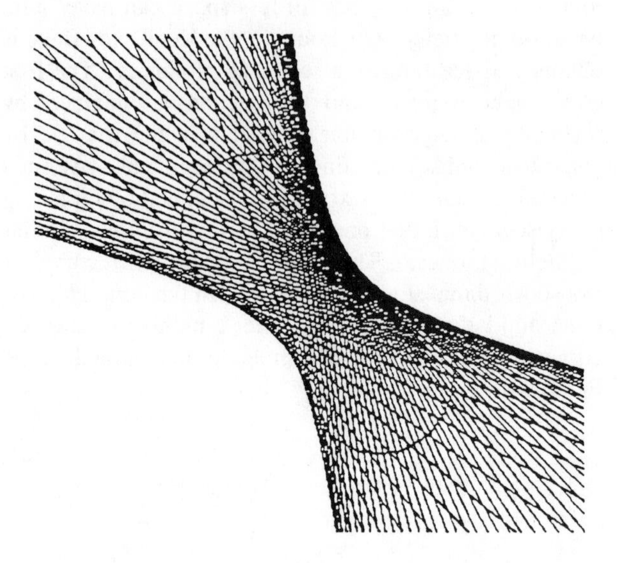

Figure 13. A hyperbola as a line conic.

Now we change the rules a bit. For example, we use one circle for marking the points, but the second time we let the turtle go twice as far for each step (and so twice around the circle). The envelope of the lines connecting these points is a cardioid (see Figure 14). If the ratio of steps is 3 to 1, a nephroid results; if 4 to 1, a three–cusped epicycloid.

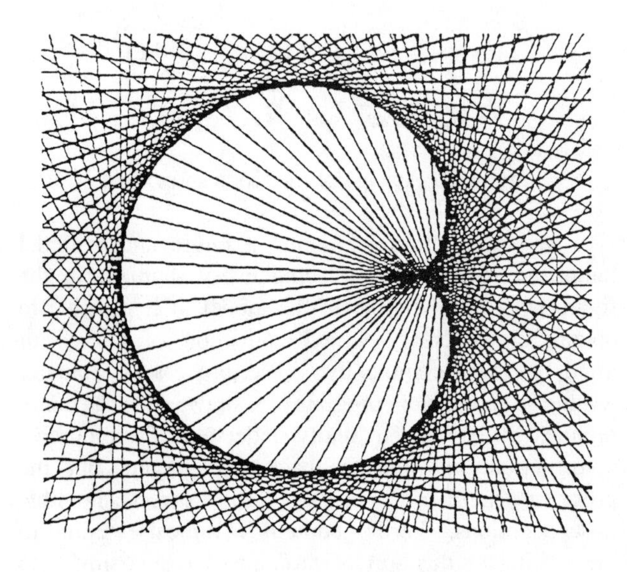

Figure 14. A cardioid as a string figure.

There are other classical constructions of line conics that are interesting to draw with computers. For instance, we choose a circle and a point P not on the circle. For every point Q on the circle, we draw the line through Q perpendicular to the segment from P to Q. Then the envelope of this family of lines is a conic. For P inside the circle an ellipse results; for P outside, a hyperbola. For more details about these envelopes, see [12]. Another book with a treatment of curves easily adapted to computers is [9].

Another interesting way to produce families of lines is to draw a curve (e.g., a spiral) made up of turtle steps and turns. After each step draw the normal line to the curve, i.e., the line normal to the direction of the turtle. For a smooth curve, the envelope of the normals is called the *evolute*. While the evolutes of some smooth curves are also smooth (e.g., the evolute of a logarithmic spiral is a similar spiral), usually the evolute has singularities. The evolute of an ellipse is an interesting example. (In the section on affine turtle geometry below we discuss one method for producing ellipses.) For a general discussion of evolutes, see [7].

When lines are interpreted as light rays, the envelope is a *caustic*, the locus where the light is focused. Look at the bright curve of light in your coffee cup; it is possible to simulate this curve by choosing a circle (more generally, an ellipse) and a point P not on the circle. For each point Q on the circle, draw the line that is the reflection of the line PQ across the tangent line of the circle at Q. (In other words, reflect the point P across the tangent line to get P' and draw the line $P'Q$.) If P is at the center of the circle (or a focus of the ellipse), the lines all pass through a single point, but otherwise they define an envelope which is the bright curve on the surface of the coffee. For the details of this, see [3]. A very thorough and accessible discussion of envelopes and related ideas is in [4].

Cycloids and mechanical devices.

Another classical method for describing curves is illustrated by the definition of a cycloid as the locus of a point on a circle which rolls along a line. In other words, if a light is attached to the rim of a rolling wheel and this is photographed by a time exposure, then the trace of the light in the picture is a cycloid. Similar definitions are given for epicycloids and hypocycloids, which are the loci of points on circles rolling outside or inside other circles. It is not difficult to simulate this type of definition with a computer. One can simply plot the position of the point using analytic geometry, but one can see the nature of the geometric construction better by having one turtle go around the first circle and another turtle go around the second circle, then taking the vector sum of the position vectors relative to the centers of the circles. (A more difficult feat would be to create an animated simulation of the rolling circle.)

This opens up another vast area of classical geometry wherein curves are defined as loci of points on various mechanical devices. Another example is the locus of a point on a ladder resting on a floor as the ladder slides down a wall (an ellipse). In this case, one can also look at the envelope defined by the segments representing the ladder (an astroid). In fact, one can recall the string constructions connecting pairs of points and plot the locus of the midpoint of each segment or the point dividing the segment in some fixed ratio. Examples of these constructions are found in [9], [12], [16]. These constructions are elementary to program. A more difficult project along the same lines would be to simulate one or more of the classical devices for performing constructions such as inversions in a circle, e.g., Peaucellier's linkage [5].

Affine geometry and home–brew turtles.

Affine geometry is the geometry of lines and parallelism without the introduction of angle and distance. Plane affine geometry is the geometry that is preserved under orthogonal projection from one plane in 3–space to another. Thus circles and ellipses are indistinguishable in affine geometry, and all triangles are equivalent, but the notion of parallelogram makes sense.

It is possible to modify the commands of turtle geometry so that drawing an ellipse is as easy as drawing a circle. The idea is to slant the plane of the turtle so that what appears on the screen is an ellipse, although the turtle thinks it is drawing a circle.

In the section on the foundations of turtle geometry, we said a turtle state was a position and a heading; an equivalent definition would be a position and a moving coordinate system, since in the oriented Euclidean plane, a coordinate system is determined by the unit vector along the first axis. We chose the unit vector on the first axis to be the turtle heading and the unit vector on the second axis by rotating the heading counterclockwise by 90 degrees. Let a state be given by (P,V,V'), where $V' = R(90)V$. Then the `FORWARD` command is given as before, by $F_t(P,V,V') = (P + tV,V,V')$, but the `LEFT` command is defined to be $L_u(P,V,V') = (P,W,W')$, where

$$W = (\cos u)\, V + (\sin u)\, V',$$

and

$$W' = (-\sin u)\, V + (\cos u)\, V',$$

according to the usual formulas for rotation. Now to slant the plane, we "deceive" the turtle by letting the turtle state be (P,H,L), where H and L are two independent vectors that need not be orthogonal or of unit length. These vectors define a moving coordinate system that is affine but need not be Euclidean. Then if L_u is defined by the same formulas,

$$H' = (\cos u)\, H + (\sin u)\, L$$

and

$$L' = (-\sin u)\, H + (\cos u)\, L,$$

a sequence of turtle commands will draw a figure as it would be projected on a plane in such a way that the original Euclidean coordinate vectors would be projected to H and L. The vectors H and L define the affine moving frame of the turtle.

Various commands other than L_u can be devised to change one such affine frame to another. The simplest would be a SETFRAME command that takes a new frame as input and changes the frame to the new one. Other useful intrinsic commands would be ones to rescale one or the other of the frame vectors by multiplying it by a scalar. It is also possible to define a FLIP command as a special case of this, with –1 as the scalar. Notice that the L vector (which is the left lateral vector by convention in the definition of the left turn L_u) can actually be on the right. In this case, all left turns will appear on the screen as right turns.

These tools can be used with a CIRCLE procedure to draw ellipses. (See Figure 15.) Similarly, they can be used with SQUARE to draw parallelograms or with TRIANGLE to draw arbitrary triangles. Mirror images can be drawn by using the same procedure twice, once before and once after the FLIP command, and shadows can be constructed with SETFRAME. The same tools can be used to illustrate which properties of Euclidean geometry are affine invariants and which are not (e.g., angle bisectors are not, medians are).

To create such a "home–brew" turtle in Logo or some other language, one must do two things. The first is to create the *logical turtle*. This means that some variable or variables are used to store the information about the turtle state, and then some procedures are defined to alter this state by the rules of F_t and L_u. As a practical matter, it may be necessary to give these commands slightly altered names so as not to interfere with the usual turtle commands (e.g., AFORWARD or AFD and ALEFT or ALT). The second stage is to use the standard turtle as a drawing tool to represent the home–brew turtle on the screen, the "screen turtle." This means that after each command to the logical turtle, the screen turtle is moved to a new position and/or heading depending on the new state of the logical turtle.

In addition to defining an affine turtle as described, home–brew turtles can be used to define multiple turtles for language implementations that come with only one turtle (or none). Multiple turtles are very convenient for constructions like the string figures, which naturally have two or more loci of points. A reference for defining turtles is [1]. A Logo implementation of the affine turtle is described in [8].

Three–dimensional geometry.

A few implementations of turtle graphics have three–dimensional geometry built in, but most do not. However, home–brew three–dimensional turtles can be defined roughly as the affine turtles were defined. The main difference is that they carry a frame of three

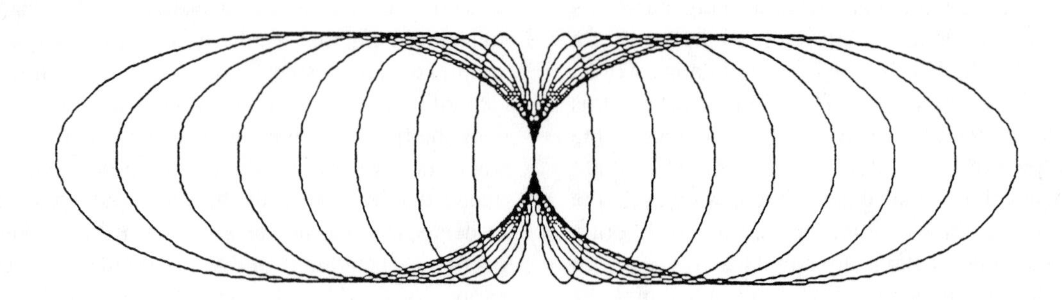

Figure 15. Ellipses drawn by CIRCLE.

vectors. (For Euclidean geometry, the vectors are orthonormal.) There are also three turning commands, one for each coordinate plane. Commanding this turtle is more like flying an airplane than driving a car.

Another difference is that the position of the logical turtle is in three–space, but the position of the screen turtle is in the plane of the screen. After each change in turtle state, the new position is projected to a new planar position for the screen turtle. This is one of the more interesting aspects of these turtles, for one has a choice of how to map three–space onto the plane of the screen.

Two natural choices are orthogonal projection and central projection. It is very interesting for students to see how a figure they have created in three–space changes with different choices of projection. How to move the figure or change the projection so that the figure looks the expected way becomes an important practical problem, and solving it provides some much needed practice with the geometry of space.

There are many things to do with this space–turtle geometry. It is interesting to draw simple figures like cubes and prisms. By drawing a family of line segments in space, one can generate ruled surfaces such as cylinders, cones, and hyperboloids (see Figure 16). One can then point out to students that the pictures on the screen are identical to some of the string figures and that there is a relationship between envelopes and projections of surfaces in space.

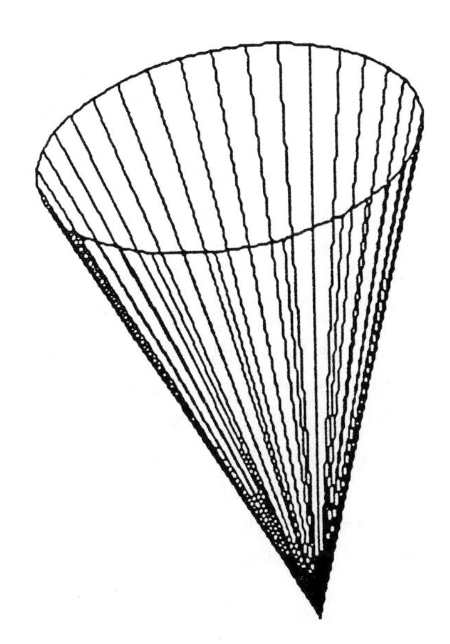

Figure 16. A cone generated by lines in space.

It is an interesting exercise to have students construct a three–dimensional scene with buildings and roads and then project it so that it looks like a typical perspective drawing with a vanishing line at infinity near the center of the screen and with the roads coming together at this infinity. This is harder than it looks, because the student has to understand how planes in the construction must be related to the projection. It gives some insight into the rudiments of projective geometry.

Some of my students have constructed the Platonic solids on the screen, with the duals in contrasting colors. Others have created aerial views of cities, three–dimensional tilings (see Figure 17), and three–dimensional fractals.

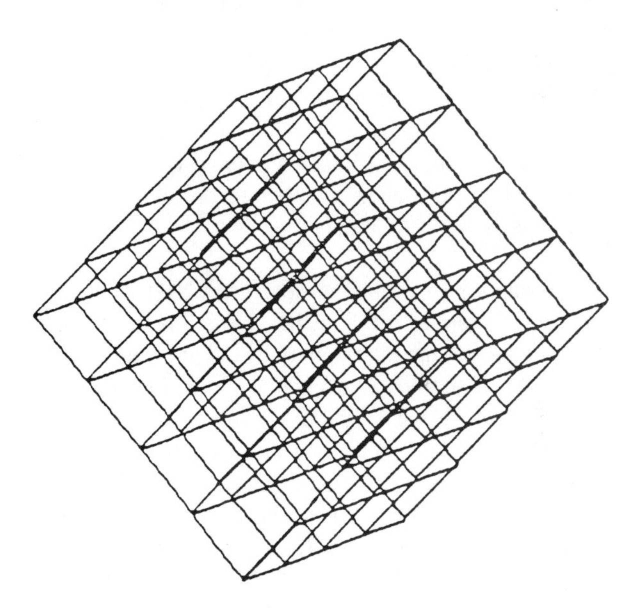

Figure 17. A three–dimensional tiling.

Although the geometry of space is very rewarding, it is definitely more difficult to explore with turtle geometry than is plane geometry. First, the geometry itself is harder. Second, it is harder to represent the objects of the geometry; for example, the wire–frame representation of surfaces has many drawbacks. Finally, home–brew three–dimensional turtle geometry is much slower than plane geometry, at least in Logo, although on the newer and more powerful machines this is less excruciating. References for three–dimensional turtle geometry are [1] and [13].

Algebra and geometry.

It is important to understand the various ways of representing mathematical objects. For example, a rotation can be represented by a vector formula, a matrix, or a complex number. It is possible to explore this circle of ideas geometrically by first defining various algebraic operations, such as complex multiplication and matrix multiplication, in a language such as Logo. Then one can plot the points generated by recursively multiplying the current turtle position (viewed as a complex number) by a fixed complex constant and view the results. If the modulus of the constant is one, then one gets a polygon; if it is not, a logarithmic spiral results. Thus one obtains the same figures studied earlier by turtle geometry, and the student can be asked to understand the relationship between the two approaches.

Students can also be asked to create procedures that perform the operations of vector algebra, such as vector sum and dot product. Then they can use these operations to write a procedure that will project one vector onto another and draw the picture at the same time. Getting the picture to match the algebra seems to be a useful exercise leading to better comprehension of what these vector operations really mean.

The equation of a line, as elementary as it is, can be used for a number of drawing experiments. First consider a procedure that simply draws the line on the screen. This actually has a few ideas in it, for one must decide in advance what kind of data about the line will be given. Students sometimes find it a challenge to take a procedure that they have written to draw a line given two points and use it to define a procedure that draws a line given the equation (where are the points?). This forces one to think about the various ways of representing lines and how one can systematically pass from one to another. This illustrates an important idea that can be difficult for students: When reasoning about an object, it is very important to take into account the representation of the object.

Some of this is a little austere for the typical student, so it is important to ask geometric questions that require him or her to confront these issues. For example, if a line is determined by its normal direction and a point P on the line, what happens to the line if the normal direction is rotated, and simultaneously the point P is rotated about the origin, perhaps at a different rate (see Figure 18). This is closely connected to the string figure geometry described above, and the rotational symmetry of the envelope is based on the same reasoning about angles as was used above in our discussion of symmetry.

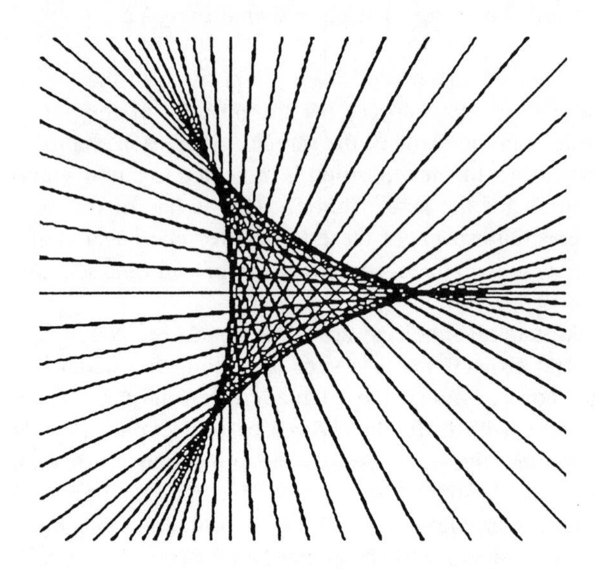

Figure 18. An envelope generated by simultaneous rotations of a point and normal direction.

Once one has a representation for lines and a means to draw them, then one can ask how they interact with turtles. For example, given a line, we may write a procedure that will move the turtle forward until it meets the line. (What to do when the turtle must move backward to meet the line, or will not meet it at all, depends on what we plan to do with this procedure.) This is a good exercise in the use of the dot product. Then we may write a procedure that will bounce the turtle off the line like a billiard ball. After this it is possible to simulate the trajectory of a billiard ball on a table of any shape.

One interesting aspect of the projects in the preceding paragraph is that they will work unaltered for three–dimensional geometry, provided they are written in terms of vector operations such as dot product and not explicitly in terms of coordinates. Thus the billiard table program becomes a racquetball program, since the equations for planes in space are the same as the equations for lines in the plane.

Other topics.

There are many interesting questions that one can raise about convexity and connectivity of regions. For example, given a finite set of points in the plane, we may write a program that will draw the convex hull of the points. Or, given a simple closed polygonal curve, we can determine whether a given point is inside or outside the curve, or create a procedure that will fill the interior with crosshatching. This sort of activity borders on the questions that are studied in linear programming and in the computational geometry used in computer graphics.

Inversive geometry, the geometry of inversion in circles, benefits greatly from computers, since examples are generated more quickly. Also, on computers with very fast arc–drawing routines, such as the Macintosh, it is possible to build a complete turtle geometry of the Poincaré disk model of hyperbolic non–Euclidean geometry that works fairly briskly. One can also consider envelopes of families of curves that are not lines, such as circles or parabolas.

Projective geometry is a very tempting subject, but it would require a careful approach to make it accessible to the beginning programmer, since figures always run off the screen. Some of the synthetic constructions of conics would be similar to the constructions of string figures and loci described above.

An ambitious but beautiful topic is the study of the curvature of surfaces and of space, which is the topic of the last half of [1]. It includes both the geometry of the sphere and the (metric) geometry of the cube. The former has constant positive curvature, while the latter has curvature concentrated only at the corners. This is explained by defining turtle geometry on these spaces and using the Gauss–Bonnet theorem to define curvature.

Conclusion.

I hope this sampling will generate some ideas about what can be done in a geometry course in which the students do extensive programming. Based on my experience, if one wants to teach such a course, the important thing is to learn a modicum of programming, read a couple of books, and then jump in. Some mistakes are inevitable, but once you get some experience with looking at geometry from the programming point of view, all sorts of ideas will present themselves. Look in old tomes for partially forgotten mathematics, and look in new books, such as [6], [10], [14], always asking the question, "Would this be interesting on computers?"

Also, look at classical geometry theorems with new eyes. For example, there is a theorem about hyperbolas that says that if a point P is on a hyperbola, the rays from P to the foci are on opposite sides of the tangent line and make equal angles with this line. Thinking about turtles, we say this: Suppose A and B are two points, and a turtle moves so that A is to the left and B is to the right, with the rays toward the two points always making the same angle with the heading. Then the turtle moves along a hyperbola. This is actually a recipe for drawing a hyperbola with a turtle. (The actual drawing will be a discrete approximation to the correct curve, which is described as a solution to the differential equation that appears implicitly in this formulation of the theorem.)

This account has concentrated on what one can teach with computers and has begged the question of why one should want to do so. There is too much to say about this to really address the question in this article, but I would like to conclude with some brief observations based on my experience. The geometry seems more immediate and alive to students; one can bring in modeling problems that show the geometry of growth in nature. The disciplined creative act of writing a program is closely akin to writing a proof, but a geometric program has the advantage that it will immediately draw a picture that will tell you whether you have the ideas right.

This immediate feedback is like having a Socratic tutor on call at all times. Students can think more like mathematicians, in that they can generate their own geometric questions and then perhaps solve them; they can also begin with simple cases and work their way up to general understanding. This is the only undergraduate mathematics course I have ever taught in which the students do original projects.

Also, it is easier for students to work at various levels of competence and expertise and still learn something and experience some success with mathematics; conventional courses are very prone to baffle the weaker students and bore the more able ones. There are the factors of surprise and serendipity: Perhaps one does something with squares and sees spirals on the screen that reveal hitherto unnoticed relationships. Finally, one additional reward for the teacher is that this geometry is new and interesting for him or her, too, and this different point of view generates fresh pleasures from geometry.

REFERENCES

1. Abelson, H., and A. diSessa. **Turtle Geometry**. Cambridge, MA: MIT Press, 1980.

2. Bezuska, S., M. Kenney, and L. Silvey. **Tesselations: The Geometry of Patterns**. Palo Alto, CA: Creative Publications, 1974.

3. Bruce, J. W., P. J. Giblin, and C. G. Gibson. "Caustics Through the Looking Glass." **The Mathematical Intelligencer 6**, No. 1 (1984), 47–58.

4. Bruce, J. W., and P. J. Giblin. **Curves and Singularities**. Cambridge, England: Cambridge University Press, 1984.

5. Coxeter, H. S. M. **Introduction to Geometry**. New York: Wiley, 1961.

6. Grünbaum, B., and G. C. Shephard. **Tilings and Patterns** New York: Freeman, 1986.

7. Guggenheimer, H. **Differential Geometry**. New York: Dover, 1977 (originally McGraw–Hill, New York, 1963).

8. King, J. "The Affine Turtle: A Turtle That Can Draw Shadows." **Logo Exchange 5**, No. 7 (1987), pp.15–16.

9. Leapfrogs. **Curves**. Cambridge, MA: Leapfrogs, 1982. (ISBN 0 905531 29 9)

10. Mandelbrot, B. **The Fractal Geometry of Nature**. San Francisco: W. H. Freeman, 1982.

11. Papert, S. **Mindstorms: Children, Computers, and Powerful Ideas**. New York: Basic Books, 1980.

12. Pedoe, D. **Geometry and the Visual Arts**. New York: Dover, 1983 (originally **Geometry and the Liberal Arts**, St. Martin's Press, New York, 1976)

13. Reggini, H. "Towards an Artisanal Use of Computers." Logo 86, The Third International Logo Conference. Cambridge, MA: MIT, 1986.

14. Stevens, P. S. **A Handbook of Regular Patterns: An Introduction to Symmetry in Two Dimensions**. Cambridge, MA: MIT Press, 1981.

15. Thornburg, D. **Discovering Apple Logo: An Invitation to the Art and Pattern of Nature**. Reading, MA: Addison–Wesley, 1983.

16. Yates, R. **Curves and their Properties**. Reston, VA: National Council of Teachers of Mathematics, 1974.

Computers in Abstract Algebra and Number Theory

Christopher H. Nevison

The advent of widespread access to computers via time–sharing or microcomputers has enabled their use in teaching in many new areas. In this essay I discuss the potential for using computers in undergraduate courses in abstract algebra and number theory and I survey some experiments that have been made. The specific experiences that are described here are certainly not all–inclusive but are based on the author's attempts to gather information over the past two years. We encourage others to share their experiences in journals and at professional meetings.

How can computers be used?

There are two primary modes of use of a computer for teaching in any area: (1) classroom demonstration and (2) student work outside class. Category (2) may be divided into two major subcategories: (a) writing programs to solve problems; (b) use of packaged software to experiment with or solve a given class of problems. In this essay I consider classroom demonstration to mean the use of software that has been written and tested to teach or demonstrate the primary subject matter. Teaching programming may be a needed adjunct to the use of the computer in some cases, but it is distinct from the teaching of the primary topic.

The use of the computer in the classroom can include efficient demonstration of examples that would be unreasonably difficult to prepare without the use of the computer. These examples might or might not be displayed with computer equipment — but the computer is the tool that has made the examples possible, or at least easier. For example, an instructor in a number theory class might develop his own examples of some congruences using a computer package with exact precision, large integer calculation.

Examples that are developed using a computer might also be enhanced with graphic displays using a computer. In this case the computer becomes essential to the example, as it is the tool that makes certain kinds of graphics possible.

Examples that are developed or displayed with computer assistance extend our teaching capabilities in only one direction and do not really alter our methods. A more exciting prospect for incorporation of the computer into the classroom is its potential for dynamic experimentation. For example, suppose we are teaching about normal subgroups and quotient groups. Computer programs exist which make it possible, with groups of a size larger than could be handled without computers, to discover a normal subgroup, find the cosets, and look at the structure of the quotient group. We can use this capability to develop and display more complex examples than might otherwise be possible and to respond to queries that we might not have anticipated, so long as they fall within the capability of the software. Thus, we have the potential to use a discovery method in our teaching more easily than has heretofore been the case.

How can computers be used by students?

(a) *Programming*. If we assume our students know how to program at an appropriate level, then there are many interesting problems that can be set as programming problems. For example, one can find all non–isomorphic Abelian groups of size k. Once the appropriate topics in an abstract algebra course have been covered, this is a problem well within the capabilities of a good programmer. Many such problems may be found, for instance, in [3] or [9].

What do students learn from writing programs such as these? There are certainly some benefits to their understanding of the concepts involved. One must know the precise content of a theorem, at least symbolically, before one can incorporate that information into a computer program. On the other hand, as anyone who has written even a modest length computer program knows, there are many other things involved with writing a program to solve a problem such as this: developing appropriate data structures, translating an abstract theorem into a concrete algorithm, translating the algorithm into code, documenting the code, testing and debugging code. While some of these, notably the high level documentation and the transition from theorem to algorithm, can greatly enhance a student's understanding of the primary topic, much of this programming process is peripheral to the main topic but still very time–consuming. Consequently, an instructor must weigh the benefits of programming problems against the costs — primarily time.

(b) *Using software packages.* Students can do a great deal in an algebra or number theory course with well–designed and developed software. There are at least two software packages that enable the user to experiment with small finite groups [9], [6]. In the first case knowledge of the language APL is required, but the user need not do a lot of programming. The second package is easily used even by a computer novice. In addition, the software package *Cayley* [1] has been designed for research investigation in algebra, but it can also be used for teaching.

With packages such as these, there are a range of problems with which students can experiment that go far beyond the traditional examples. A well–designed package can enable students to experiment with a variety of examples and thereby develop a better understanding of definitions of properties (e.g., left–identity, right–inverse, coset, etc.) and of theorems. Students can investigate their own conjectures. A package for exact precision, long integer arithmetic [including greatest common divisor and $r = b^a \pmod{n}$] can enable students to verify theorems of number theory and try their own conjectures.

In abstract algebra and number theory, the capability of the computer to do large examples in a reasonable time extends our previous capabilities to the point where there can be a qualitative change in our methods. The potential for meaningful experimentation requiring computation well beyond the usual simple examples makes an experimental discovery method feasible.

The use of software packages has drawbacks as well. Instructors must consider the shift in emphasis that their use necessarily implies. Are more and more complex examples the right way to investigate a particular topic? Should we emphasize those topics that are amenable to computer calculations over those for which the computer cannot be used or for which software is not available? As with any effective tool, the use of the computer must be considered in the context of the primary objectives of the course.

The investigation of applications to computers and computer science.

Applications of algebra or number theory to computers and computer science fall more in the category of "what we teach" (curriculum), rather than "how we teach it" (methodology), which is my main concern in this essay. Nevertheless, applications can be used as wonderful motivation when teaching many subjects, and algebra and number theory are no exceptions. When these applications can also be demonstrated effectively with computer technology, the "what" is certainly overlapping with the "how."

In that spirit, I offer two examples of applications that are excellent motivation for certain topics in algebra and number theory, and that have the potential for computer demonstration or experimentation: Algebraic analysis of error–correcting codes, such as Hamming codes, illustrates some interesting ideas in group theory [3]. And public key cryptography is an important application that depends critically on number theory and can be demonstrated effectively with a computer. Topics such as these and the use of computers to demonstrate them should find their way into our courses in algebra and number theory.

The use of computers in teaching abstract algebra: some examples.

In this section we will describe three instances where computers have been and are being used to teach abstract algebra. At this point, there have been only a few such experiments and there are only a limited number of software packages available. Perhaps these examples will encourage others to undertake similar experiments or develop suitable software.

Charles Sims at Rutgers University has developed over the past several years an abstract algebra class that includes computer analysis as an integral part. The materials he has developed are available as a textbook [9] and additional supplementary materials, including a package of APL programs called CLASSLIB (see the book for further details). Sims chose APL because, in his own words [9]:

1. APL is implemented in an interactive mode.

2. Arrays exist independent of programs.

3. One–line statements can be entered and executed immediately. In effect, beginning students do not have to write programs in the traditional sense.

4. The language contains many powerful primitive operations for manipulating arrays that are very useful in describing algebraic algorithms.

Despite using a computer language, the objective of Sims' course and textbook is to teach traditional abstract algebra, not applications. The computer is used as a tool to investigate examples beyond reasonable hand calculation. For example, here are a few of the exercises from the book, selected at random:

> Write an APL proposition corresponding to the assertion that the inverse of the product of two elements of (*G*6,*G*6) is the product of the inverses in the opposite order. [9, pp. 75–76]

> Use GPSGP [an APL program from CLASSLIB] to find the orders of the groups generated by
> (a) (0) (1,2,3,4,5) and (0,1) (2) (3,4) (5).
> (b) (0,1,3,2,5,6,4) and (0) (1) (2) (3,4) (5,6).
> [9, p. 124]

> Construct addition and multiplication tables for Z_6. Set $N \leftarrow 6$, and set the current finite ring to Z_6 using FRINT [a CLASSLIB program]. Compare the speeds of execution of the procedures with the prefixes *ZN* and *FR* for some sample calculations with arrays over Z_6. [9, p. 166]

Of course, many traditional problems that do not use the computer are also included.

Sims has found that the facility for experimenting on examples with the computer helps students develop a deeper understanding of the concepts. In addition, when students program the computer to carry out algebraic algorithms, they develop a different perspective from one based solely on theorems about algebraic facts.

Some instructors may feel the need for learning a computer language is a drawback for this approach. It certainly requires more time, as Sims recognizes: "The time required to introduce students to APL is too great to leave sufficient time to cover a reasonable amount of algebra in a one–semester course ... Approximately three weeks are needed to cover this material [the APL programming introduction]." [9, p. viii] Consequently, Sims recommends a year–long course. On the other hand, such a course enables students to see traditional abstract algebra from a new and increasingly important perspective.

Joseph Gallian at the University of Minnesota at Duluth has also experimented with the use of the computer in abstract algebra. He reports: "I have used the computer quite extensively in connection with abstract algebra but not in a formal way ... The computer exercises in my book are never assigned as homework, although a handful of students have done some of them ... These students inform me that it does help them understand the ideas involved." [4] Nevertheless, Gallian reports several successful student projects that have developed from his abstract algebra course and the computer applications he has discussed. Some of these are described in [2].

Gallian has included many computer problems in his textbook [3]. These require the students to have a knowledge of programming. However, in a context where students can program, as at many colleges where at least one programming course is required, some of these problems could become interesting assignments for an abstract algebra class. Some examples, randomly selected, will give the flavor of Gallian's computer related problems:

> "Write a program to print out the cyclic subgroups of $U(n)$ [the group of all positive integers less than n and relatively prime to n, under multiplication modulo n], generated by each K in $U(n)$. Assume $n < 100$. Run the program for $n = 12, 15, 30$. Compare the orders of the subgroups with the order of the group itself. What arithmetical relationship do these integers have?" [3, p. 58]

> "Write a program to list all the subgroups of Z_n. Run your program for $n = 8, 9, 10, 12$, and 30." [3, p. 69]

> "Write a program that will print out $U(n)$ as an internal direct product as indicated in Theorem 9.2. Assume the relatively prime factors of n are given by the user. Test your program for $105 = 15 \cdot 7$, $105 = 3 \cdot 5 \cdot 7$, $105 = 35 \cdot 3$, $72 = 9 \cdot 8$, $360 = 40$, $360 = 45 \cdot 8$." [3, p. 116]

> "Let $Z(n)[i] = \{a + bi \mid a, b \in Z(n), i^2 = -1\}$ (the Gaussian integers modulo n). Write a program to find the group of units of this ring and the order of each element of the group. Run your program for $n = 3, 7, 11$, and 23. Is the group of units cyclic for these cases? Try to guess a formula for the order of the group of units of $Z(n)[i]$ as a function of n when n is prime and n = 3 modulo 4. Run your program for n = 9. Does your formula predict the correct order in this case? ..." [3, p. 173]

"Program the algorithm given in exercise 30 to find the irreducible factorization over **Z** of all polynomials of the form $X^n - 1$, where n is between 2 and 100. On the basis of this information, make a conjecture about the coefficients of the irreducible factors of $X^n - 1$ for all n. Test your conjecture for n = 105." [3, p. 223]

"Program the Index Theorem. Use a counter M to keep track of how many integers it eliminates (as candidates for the order of a simple group) on any given interval. Run your program for the intervals 1–100; 501–600; 5001–5100; 10,001–10,100. How does M seem to behave as the sizes of the integers grow?" [3, p. 305]

Although the time is here when we can reasonably assign problems like this to students in an abstract algebra course, and no longer simply suggest them as extra work, the time needed to write a program for such problems precludes the use of more than a couple in a course. The utility of having students do such programs is twofold: They gain deeper understanding by working the algorithm into a running program, and they can experiment with their own conjectures, based on results from these programs, that go well beyond what is possible with hand calculation.

The latter benefit can also be obtained if suitable "canned" programs are available. Ladnor Geissinger at the University of North Carolina at Chapel Hill has developed a package of programs that enable the user to experiment with small groups [6]. Geissinger recommends the use of his package for both classroom demonstration and student experimentation:

"The instructor can use this program to demonstrate to a class many things which are not possible or too time–consuming using just a blackboard ... All the conjugates of an element (or subgroup) can be easily displayed, and the role of the centralizer (center) and its cosets can be clearly seen. Students always have trouble with these things and need to see lots of nontrivial examples worked out and explained.

"[Students] can explore on their own a wide variety of operations and gradually gain a feeling of hands–on experience with some abstract algebraic structures." [6].

Geissinger has used preliminary versions of this package in his own classes and reports that it gives students the chance to experiment with many more examples than by hand:

"There are lots of exercises which I can now give for them to do using the program which I previously would have considered unthinkable — too large, too time–consuming, too complicated. One of the benefits which I hadn't given much thought to before designing the program is the repeated exposure to the basic definitions. I put in brief definitions which appear just before each of the different calculations is carried out, and students have frequently commented that this was very useful to them." [5].

Geissinger's package enables the user to experiment with groups and other structures with 3 to 16 elements. These are represented by their operation tables, and the program has the facility to check various properties, such as commutativity, associativity, and left, right, and complete identities and inverses. The program has the tables for all 40 non–isomorphic groups in this range, as well as many other structures. The user can enter his own operation table and run checks on it. The program is very user friendly and really needs no external documentation, as it includes appropriate instructions in its displays. It runs on an IBM PC; a color monitor is preferred but not absolutely required.

John Cannon and Jim Richardson at the University of Sydney, Australia, have developed the Cayley system for use in research in algebra. They have used this system for instructional purposes with some success [1]; the following summary is from that paper:

The programs share a common overall approach, which may be summarized as follows:

(a) Each program provides the student with a "calculator" for dealing with the mathematical objects appropriate to the area of study: matrices and vectors in MATRIX, groups in Cayley. Commands are provided at a variety of levels, so that the calculator can grow in power as the student learns.

(b) For a given course, a file of problems is provided by the instructor. This may contain both human–language text for the student, giving exercises and hints, and commands for the program, to set up the required data.

The major aims of the programs are to remove the burden of arithmetic or other routine calculation from the student, thus allowing larger, more realistic, and more numerous exercises to be attempted, and to deal with common student difficulties through the hints.

Cayley is designed primarily as a research tool and is essentially a complete programming language for algebraic computations. Cannon and Richardson suggest that it can be easily used by students for calculation, although they do recognize some drawbacks:

> Since Cayley was designed principally as a research tool, there are one or two aspects of the program's present form which can be troublesome in teaching conditions. Students often ask for the provision of a command–line editor, which would enable them to correct minor mistakes in a long command without typing it all again from scratch. To be usable by all students, including those with little prior familiarity with computers, such an editor would need to be very simple to operate: something along the lines of the editing features commonly available in the form of "arrow" cursor–motion keys on microcomputers is desirable, but this would be difficult to implement in Cayley's present context.
>
> Cayley can use substantial amounts of CPU time: even if students work only with groups of low order, an ill–chosen iteration can be expensive. It is possible to impose a time limit on an entire Cayley session, but it would be worthwhile to add an internal limit to interrupt excessively time–consuming commands without losing all work already carried out.

Cayley is a large software package and therefore is available only on large mini– or mainframe computers. Perhaps in the future a similar but less general tool will be developed for microcomputers. In summary, Cayley is a very useful tool for research that certainly can be applied to the teaching of algebra. If it is available, it is worth experimenting with for teaching purposes.

The experiences described above represent a range of ways to use the computer as a tool for teaching abstract algebra. The first two approaches, which involve programming at some level, can have tremendous benefits, but at relatively high costs in terms of course time and student time. The software packages for experimentation in abstract algebra are perhaps the most promising trend for the future. They show how tools can be provided to extend the range of problems that students can explore *without* requiring inordinate amounts of time to be spent on peripheral matters. Certainly much remains to be done so that the problems suggested as programming problems by Sims and Gallian come within the range of software packages as easy to use as the one developed by Geissinger.

Some prospects for using computers in number theory.

There are two textbooks that develop topics in number theory that are amenable to the use of computers [8], [7]. Although these books are not necessarily designed for including regular computer use in a number theory course, they are certainly suggestive for those with interests in this direction.

Schroeder's book [8] is an exploration of applications of number theory in many fields. Thus it is not a number theory textbook, but rather a book that, in the words of the author,

> "seeks to fill the gap in basic literacy in number theory ... in a manner that stresses intuition and interrelationships, as well as applications in physics, biology, computer science, digital communication, cryptology and more playful things, such as puzzles, teasers and artistic designs." [8, p. viii]

It includes topics from computer science, such as public–key cryptography and random number generation, that can be used to motivate various topics in number theory. Schroeder also includes some programs for a programmable calculator that can easily be adapted to computer programs. These are potent tools for investigating examples, just as the programs for algebra mentioned above are.

Riesel's book [7] is also not a text for a standard number theory course. It investigates the properties of primes and leads up to a thorough discussion of public–key cryptography. In addition, it has many programs for the investigation of the phenomena discussed. The Pascal listings can be easily adapted to several computers (although the code needs some reworking to take advantage of the reliable structure of Pascal; the current versions look like a direct translation from FORTRAN) and used as tools for experimentation with many ideas in number theory. Among the programs included are these:

A package for long integer exact arithmetic (which can be done for quite large integers, depending on the capability of the computer), including addition, subtraction, multiplication, division with remainder, greatest common divisor, exponentiation modulo a modulus;

Fermat's test for pseudoprimes;

A strong pseudoprime test;

Trial division (for factoring);

Pollard's factor searching algorithm.

In addition, there are several suggestions to the reader for other programs and some discussion of the issue of computational complexity for problems like factorization. Riesel's book can be a useful source for instructors of number theory courses who want to make use of the computer.

John McCleary of Vassar College has under development another project using the computer in number theory. He plans to produce a short book suitable for an enrichment course, as might be offered to talented high school students or for a college freshman seminar, covering the number theory needed to develop the RSA public key coding scheme. The book will be packaged with computer programs that I have developed; these provide the user with calculator–like tools to do large integer arithmetic, including finding greatest common divisors and exponentiation modulo a modulus. These materials worked very successfully in an initial trial with a group of talented high school students in the summer of 1986.

Summary.

The real potential for the use of computers to enhance teaching in theoretical fields such as abstract algebra and number theory is only now beginning to be realized. The fact that we found only the few examples cited here, although there are undoubtedly others, demonstrates the scarcity of materials for using the computer in these courses. On the other hand, our examples show that it can be done. As individuals develop good software packages and the emphasis moves away from programming or the use of a specialized language toward the use of software for experimentation, we will find more and more opportunities for enhancing our teaching in these areas.

I wish to thank those who have contributed information on their experiences for the development of this essay, and I encourage others who have worked with computers in these courses to share their ideas.

REFERENCES

1. Cannon, John, and Jim Richarson. "Cayley — Teaching Group Theory by Computer." **SIGSAM Bulletin 19**, no. 1 (February, 1985), 15–18.

2. Gallian, Joseph A. "Computers in Group Theory." **Mathematics Magazine 49** (1976), 69–73.

3. Gallian, Joseph A. **Contemporary Abstract Algebra**. Lexington, MA: D. C. Heath, 1986.

4. Gallian, Joseph A. Private communication, 1986.

5. Geissinger, Ladnor. Private communication, 1986.

6. Geissinger, Ladnor. **Exploring Small Groups** (a software package with documentation). Saddle Brook, NJ: Harcourt Brace Jovanovich, 1987.

7. Riesel, Hans. **Prime Number and Computer Methods for Factorization**. Boston: Birkhauser, 1985.

8. Schroeder, M. R. **Number Theory in Science and Communication**. New York: Springer–Verlag, 1984.

9. Sims, Charles C. **Abstract Algebra, A Computational Approach**. New York: John Wiley & Sons, 1984.

Computers in the Remedial Mathematics Curriculum

Gloria Gilmer and Sheldon P. Gordon

Remedial mathematics courses account for almost 47% of all enrollments in the mathematical and computer sciences in two–year colleges and about 15.5% of enrollments in mathematics in four year colleges. Two–year college faculties cited remediation as their biggest problem in the mid–1980's, and universities and four–year institutions still regard remediation as a major problem, if not a growing one [1].

At present, almost all remediation efforts in mathematics involve repetition of high school mathematics, and even the format remains the same. Students are presented structured tasks that lead them to follow fixed paths toward predetermined goals. They spend countless hours trying to master numerical and symbolic procedures, most of which have now been programmed into calculators. Such courses offer little promise of developing the style of reasoning that characterizes better mathematics students. Students rarely encounter open ended questions that allow them the freedom to decide how to interpret the question and how to approach it, nor do they encounter situations that involve an open search or that require more reasoning ability than memory. This has led mathematics educators to describe current approaches to remediation as inadequate and to call for alternative approaches. In particular, they have singled out the need for appropriate curriculum materials for the target population [2].

The promise of technology.

The integration of computers into the remedial mathematics curriculum offers an opportunity to address and alleviate the mathematical difficulties in such programs today. In a computer environment, students may practice skills and review concepts and principles in interesting formats (see the paper by Wenger elsewhere in this volume). They may gain experience in applying mathematics to computer simulations of real life situations. They can solve more complex mathematical problems than would be possible without the computer. They can be encouraged to use inductive approaches that motivate mathematical conjecturing. And they can explore mathematical concepts by means of graphics, thereby sharpening their intuition and learning to enjoy the experimental side of mathematics as well. Computer environments can provide individuality in the ways students come to know mathematics — an approach that is strikingly different from the traditional one. All of this is possible, but none of it has become standard practice.

Our survey of present practice.

During the summer and fall of 1986, we gathered information on remedial mathematics courses at the college level from a variety of sources, including telephone interviews with educators at fourteen institutions. Half of the respondents were from institutions that had been determined in 1985 to have exemplary developmental programs by the National Center for Developmental Education at Appalachian State University in Boone, North Carolina. The remaining respondents were individuals who had published papers or spoken at conferences about remediation. A list of the respondents and their institutions appears at the end of this paper.

Our plan was to find those characteristics of remedial programs that use computers. We asked for the following information:

- titles of courses in which computers are used most often;
- purposes of computer use;
- instructional settings;
- numbers of students served and average time spent by them on the computer;
- students' attitudes toward computer use in these classes;
- topics treated and with what success;
- most popular hardware configurations and software packages;
- problems encountered; and
- predictions for computer use in these courses over the next five years.

The rest of this article reports the results of this survey with occasional statements of our own opinions.

Patterns of computer use.

Our respondents reported using computers largely for drill and practice on manipulative skills in arithmetic and beginning and intermediate algebra courses. In a few instances, tutorials were in use. The numbers of computer–using students ranged from 50 to 650 per term; the larger numbers were generally in programs that *required* computer use.

Computer–assisted instruction is typically offered through a centralized learning facility or through a departmental computer facility. Centralized facilities are usually better funded with respect to staffing, equipment, maintenance, and other associated expenses. There is great variation in instructional delivery systems between those schools that have centralized remedial or developmental programs and those in which individual academic departments manage remedial instruction. In the former, computer use for mathematics is almost exclusively on a drop–in basis, with little or no direct connection to formal mathematics courses. In the latter, greater coordination exists. For example, computers may be structured into existing courses and students take a remedial course in a computer laboratory, as is the case at North Carolina A & T University. At some schools, students are required to spend a certain number of hours in a computer lab to supplement course work. For example, at the University of Akron students must spend three hours per week at the computer, and at North Carolina A & T University two hours per week are required. The most common structure for the delivery of instruction seems to be a computer laboratory within a mathematics department, where student use is determined by faculty members on an individual basis. Often one faculty member is in charge of the laboratory and is assisted by student tutors or technical assistants. Some of these labs are run entirely by tutors.

At most of the institutions in our survey, use of the computer laboratory is optional, not required. Examples of this pattern may be found at the University of Wisconsin at Whitewater, Clinton Community College, and Indiana Vocational Technical College. In some cases, computer use is not departmentalized but is the province of one or more instructors who either recommend or require use of the computer.

Finally, at Stevens Institute of Technology, where each student must own a DEC PRO 350 computer, students are notified of their mathematical deficiencies, presented with a list of books, printed materials, or actual computer disks, and instructed to correct the deficiencies in any way they choose within a given time frame.

In formal developmental skills programs, computer laboratories are one of a variety of instructional supports that might be provided. For example, tutoring services and A/V equipment are also frequently used. Such programs are more likely than departmental ones to be run by full–time staff, since they are usually multi–disciplinary. The developmental skills centers tend to be more alert to the need for conducting studies of academic success based on different methods of instruction than are programs administered by mathematics departments.

Hardware.

Our study indicates that the Apple IIe is still the most popular microcomputer used in these programs, with the IBM PC well behind. The trend for new acquisitions, especially for large scale centralized computer laboratories, seems to favor the IBM PC and its clones [11]. At this time, there is little difference in the commercial software for the two families of computers. However, several respondents indicated that the most powerful authoring systems seem to be those designed for the IBM, and they suggest that this may give the IBM PC a greater edge in the future, as more sophisticated CAI packages become available. Less commercial software is available for other computers. Thus, where such computers have been acquired, there is a pressing need to produce software locally.

Software.

Most respondents reported the use of both commercial software packages and locally developed packages. The commercial package most often cited was Eduware's *Algebra I, II,* and *III.* Frequently, the non–commercial software used was obtained from grant–supported projects at other institutions.

There is a much more limited selection of software in departmental than in centralized facilities. Where faculty members have personally produced some of the software, they seem to have a greater level of satisfaction with the software. While this can be attributed to the fact that they have a personal interest in the software, it is also true that an instructor's enthusiasm can be infectious and can be a very positive force in getting students to use the programs.

The software used in formal computer laboratories in developmental skills programs tends to be primarily commercial. While we would ordinarily expect such products to be superior to home–grown programs, the staff in these centers are often less than enthusiastic about the software they use. References [3] and [4] provide a database of college level software materials that support mathematics instruction; this includes software packages of interest in remediation programs.

Modes of computer use.

The most prevalent mode of computer use in remedial areas is computer assisted instruction (CAI) used to support drill and practice of basic arithmetic and algebraic skills. Interactive computerized tutorials are also becoming widely used. Such software provides a variety of instructional activities geared specifically to the individual needs and pace of the students. The computer's capacity to continually supply repetitious examples, problems, and explanations matches the needs of many students who otherwise could not get sufficient individual attention from an instructor. The ability to provide real–time checking of all numbers or letters typed in and to give appropriate and immediate error messages can help students learn to overcome their mistakes. More sophisticated software also provides the appropriate interactive explanations and examples to further increase student understanding and skills. Moreover, some students benefit from the impersonality of the computer; it is less embarrassing for their work to be monitored by the computer than by a person looking over their shoulders.

By removing the need to write out all the steps, some CAI software, such as *Graphing Equations* by Dugdale and Kibbey (CONDUIT), allows slower students to concentrate on the mathematical ideas without getting lost in the computations. Presumably, this leads students to a greater understanding of the mathematical methods and ideas and thus to improvement in their ability to perform such operations. The computer is particularly ideal for those students who are returning to education after a long hiatus and for whom a conventional remediation class is excruciatingly slow.

However, our respondents feel that the computer is not particulary suited for conveying new concepts, basic principles, theory, proofs, or word problems to their students. They find that most students at this level have difficulties working through text materials on their own. Thus, presenting text on the screen is often no more effective than presenting it in books. Their experiences indicate that the most effective communication of abstract ideas takes place with a live instructor.

The maturation of the software industry may change this notion to some extent. For example, the *Geometric Supposer* series (Sunburst) on circles, triangles, and quadrilaterals allows students to make geometric constructions on one of these shapes and obtain related numerical data to form the basis for conjecture and discovery in a style consistent with that of professional mathematicians. The measurement capabilities of the program encourage students to construct many examples and so develop the intuition needed for a mathematical proof [7]. This software incorporates visual methods with abstraction and generalization, thereby dispelling the notion that success in mathematics depends upon memorization of rules and formulas. This series may serve as a prototype for future software packages dedicated to discovery of mathematical proofs.

Most of our respondents indicated that students appear to enjoy the use of computers in their courses. At one institution, however, students complained that it was a waste of time. Clearly, student satisfaction is based upon a variety of factors, not the least of which is the quality of the software. Most respondents said that computer use helped the academic status of many students in remedial programs, though not all students gain from such efforts.

Problems encountered.

One of the major problems limiting the educational use of computers is the lack of suitable and portable software. Prior to 1986, 17 colleges and universities decided to require students to buy microcomputers, and another seven strongly recommended they do so, but no additional institutions made such a decision in 1986 [9]. The software problem was seen as one of the main reasons for this. This problem is seen as one that will not be solved until computer manufacturers and software developers establish a standard for operating systems that will allow software to run on all computers.

Even where computer availability is adequate, effective strategies are still lacking for integrating the computer into remedial courses. It is necessary to match computer capability with the needs of students and to fit the delivery of services into the institution's organizational structure. For example, one respondent complained that class time was interfering with the required lab time, and therefore students tended to decrease their time for "hands–on" activities. This confounded the observed effects of computer use on achievement.

Probably the greatest problem encountered in implementing any of these efforts is lack of funds. Very few institutions feel they have adequate computer facilities to meet student needs. In fact, the only respondents who felt their facilities were adequate were those whose institutions require all entering students to purchase their own computers. Obtaining the funds required for needed equipment is a major problem at many colleges and universities. Furthermore, additional equipment invariably requires additional space, and this too is a severely limited commodity on most campuses. As the facilities improve, there is a need for more formal staffing of computer laboratories. They can no longer be managed on an ad hoc basis by one or more faculty members who contribute their own free time.

The presence of computers necessitates a change in outlook both within a mathematics department and in the institution's administration. The presence of chalk and a blackboard is no longer adequate to support mathematics instruction. Maintenance of equipment and trained technical personnel may be quite costly and may require funds that are not available. Levin and Meistro [5] have reported that less than 11% of the cost associated with CAI is attributable to the hardware. The balance, almost 90%, is attributable to personnel, software, training, and other factors.

Vargas [10] suggests that many CAI programs contain serious instructional flaws. Research shows students do not learn information presented on a screen unless they are asked to respond to it. Yet some current tutorial software presents multiple screens of rules and examples (essentially a textbook on the screen) before requiring a response from the student. However, without student responses, a program cannot adjust to the student's level and cannot give feedback. Without feedback, students are uncertain about their actual understanding, which often leads to inefficient learning strategies and frustration. Vargas also notes that the way student responses are entered in many programs robs students of instructional time. This is significant, as programs that encourage rapid responses have been shown to improve achievement.

Schools that have opted for computers other than Apples and IBM PC's (or their MS–DOS clones) feel the lack of outside software and a concomitant need to produce their own. This requires a commitment on the part of the institution to provide funding and released time for faculty to devote themselves to such efforts. Software for remediation is much harder to write than either numerical or graphics application packages for higher level courses, even with the good authoring systems now available. CAI materials are considerably longer and more detailed than application packages because they have to anticipate and react appropriately to every probable student response, and they must address a wider set of topics. Our respondents feel that there is relatively little good tutorial software commercially available. Several noted that software publishing is far from being as standardized as textbook publishing.

Future prospects.

Some of our respondents see mathematics departments becoming more involved in remedial programs in the next several years. Most envision more sophisticated tutorials that go beyond simple drill and practice. Many see more creative use of the graphics capabilities of computers as a means to bring a new geometric perspective to traditionally non–geometric topics and courses at the remedial level. In particular, Logo is being used to change the focus of plane geometry from a dry repetition of old facts to an exciting discovery learning process for students; it can also be used to increase the geometric content of remedial courses to include topics from solid geometry, analytic geometry, transformational geometry, and even non–Euclidean geometries. (See the papers by Jones and King in this volume for related ideas on ways to use Logo.)

The greatest future impact may be the influence of computer algebra systems (such as MACSYMA, Maple, muMath, REDUCE, and SMP) on remedial instruction. Such systems allow the user to define an expression, apply an operation, and manipulate the output; they are easy to use and require no programming effort by the student (see [8] and the article by Hosack elsewhere in this volume).

Here are two examples that are relevant to the way we teach algebra. An expansion operator can expand $(x + y)$^4 and provide the output:

$$x^4 + 4x^3y + 6x^2y^2 + 4xy^3 + y^4.$$

(Whether displayed in this form or all on one line depends on the system, but even muMath can now print proper exponents when enhanced with the recently announced CALC–87 package — see Newsletter 16 from The Soft Warehouse.) The muMath (or other CAS) command

SOLVE (x^2 + 6x + 8 == 0, x);

quickly gives both solutions of the equation $x^2 + 6x + 8 = 0$. (The "== 0" part of the expression is optional in most CAS's.)

Now symbolic computation has been combined with numerical computation in a hand–held calculator (blurring the distinction between "calculator" and "computer"), the Hewlett–Packard 28C. The quadratic equation problem can be given to the HP 28C by entering 'x^2 + 6*x + 8' and 'x', then pressing the SOLV and QUAD keys. The answer appears as '(–6+s1*2)/2', where s1 represents ± 1. (The variable s1 can be assigned either of its values either before or after solving the equation. If before, the answer appears in numeric form; if after, an EVAL key gives the corresponding numeric answer.) The expansion problem suggested above can also be done on the HP 28C, but it is done with the TAYLR key, a connection that may not be obvious to algebra students. This calculator (which can also do symbolic calculus and matrix algebra) is certain to become popular. It lists for under $250, and a printer is available for approximately $135. Both calculator and printer will also graph functions.

As computer algebra systems become more user–friendly, their effects are likely to force the mathematics community to reassess the need for instructional emphasis on many traditional manipulative skills. The CASE NEWSLETTER, published by the Mathematics Department at Colby College, is a good source for current information and ideas on these systems.

Conclusion.

A great many institutions have incorporated the computer into their remedial mathematics programs with motivations made up partly of hope and partly of desperation. They face a seemingly insurmountable problem in bringing large numbers of students to a level at which they can learn college–level mathematics and use it in a variety of disciplines. Most of these efforts have been successful with certain students, though rarely with all. However, the results have not come close to the high expectations that some had when they began these efforts. Consequently, many of those who jumped into computer projects early now look forward to the next generation of software (as anticipated in [7]), which will allow for conjecturing, questioning, imagining, examining numerous approaches and possible outcomes, and recognizing that the right answer in one context might be inadequate in another. This approach enables students to take control of their own learning. They will find that questioning is the starting point from which they can begin to learn. They will learn how to tackle a question by representing the problem in various ways, considering special cases, searching for a pattern, and generalizing. Both high school students and adults re–entering education adjust rapidly to this style of learning. The satisfaction and pleasure derived provide motivation for future learning, which is, after all, the primary goal of remediation.

Acknowledgements.

The authors wish to express their sincere appreciation to the following individuals who provided the assistance and information on which this article is based: Henry Africk (New York City Technical College), Geoffrey Akst (Borough of Manhattan Community College), Richard Anderson (Louisiana State University), Julia Beyeler and Wayne General (University of Akron), John Butcher (Kean College of New Jersey), Michael Clippinger (Indiana Vocational Technical College), Roberta Dees (Purdue University – Calumet), Deborah Harris (North Carolina A & T University), Joseph Howland (St. Petersburg Community College), Larry Levine (Stevens Institute of Technology), Marilyn Lyons (Clinton Community College), Anita McDonald (University of Missouri – St. Louis), John Miller (City College of New York), Christopher Nevison (Colgate University), Don Norris (Ohio University), Elmer Redford (University of Wisconsin at Whitewater), Don Small (Colby College), Ronald Wenger (University of Delaware), Richard Widder (MAA), and Ed Zeidman (Essex Community College).

REFERENCES

1. Anderson, R. D., and Donald Albers. **Programs in Mathematics and Computer Sciences, 1985–1986**. Washington, DC: Conference Board of the Mathematical Sciences, 1986.

2. Conference Board of the Mathematical Sciences. **New Goals for Mathematical Sciences Education**. Washington, DC: CBMS, 1983.

3. Cunningham, R. S., and David A. Smith. "A Mathematics Software Database." **The College Mathematics Journal 17** (1986), 255–266.

4. Cunningham, R. S., and David A. Smith. "A Mathematics Software Database Update." **The College Mathematics Journal 18** (1987), 242–247.

5. Levin, H. M., and G. Meistro. "Is CAI Cost–Effective?" **Phi Delta Kappan 67** (1986), 745–749.

6. Mura, Roberta. "Feminist Views on Mathematics." **Association of Women in Mathematics Newsletter 17** (Jul.–Aug., 1987).

7. Schwartz, Judah, and M. Yerushalmy. "The Geometric Supposer: An Intellectual Prosthesis for Making Conjectures." **The College Mathematics Journal 18** (1987), 58–65.

8. Small, Don, John Hosack, and Kenneth Lane. "Computer Algebra Systems in Undergraduate Instruction." **The College Mathematics Journal 17** (1986), 423–433.

9. Turner, J. A. "Drive to Require Students to Buy Computers Slows." **The Chronicle of Higher Education 33** (Feb. 4, 1987), 1, 28.

10. Vargas, J. S. "Instructional Flaws in Computer Assisted Instruction." **Phi Delta Kappan 67** (1986), 738–744.

11. Warlick, C. H. (ed.) **Directory of Computing Facilities in Higher Education**. Austin: The University of Texas at Austin Computation Center, 1986.

Computers in "Transition" Mathematics Courses: Pragmatic Experience and Future Perspectives

R. H. Wenger

The Mathematical Sciences Teaching and Learning Center at the University of Delaware has the responsibility for teaching three courses considered to be "transitional" because of their overlap in content with high school courses. During our six years experience teaching 2500 students per year in these courses, we have used both a sophisticated mainframe instructional system and microcomputer systems in both self–paced and traditional versions of each course.

In this paper I discuss some highlights of this experience. I also identify some of the learning issues for which new computing environments might be designed. This paper is *not* a survey of the projects in this area. Rather, I hope that one institution's experiments and experience in supporting *full courses* for large populations of students at the introductory level will provide some insight into the value and complexities of such efforts. The primary themes of this paper are these:

1. The limitations on development of more effective uses of computers in transition courses are *not* the lack of computing equipment or software but rather the lack of sufficiently clear and principled descriptions of the forms of understanding desired by mathematicians, educators, and textbook authors.

2. Among the most powerful advantages of interactive computer technologies over text–based materials are:

- *redundancy*, i.e., focusing and *re*focusing attention of the learner on the salient features of a concept or mathematical object to support the development of subtle inferences, abstractions, and generalizations;

- influencing the *organization* of the learner's mathematical knowledge, i.e., associating concepts, representations, procedures, and their uses, and effectively assimilating new concepts and structures with old ones;

- providing experience that allows the learner to associate *abstract* concepts and structures with more *concrete* referents;

- influencing the *beliefs* of students about the meaning of "doing mathematics," that it is not just "memorize, match, and mimic" activity, but rather an exploratory, creative enterprise.

3. Efforts to use computers to enhance the learning of mathematics are a *critical catalyst* for clarification of the purposes of the curriculum and the related instructional strategies and materials. Furthermore, this effect on mathematics education may be significantly more important than the actual uses of computers in the classroom, at least for the next decade.

The remainder of this paper may be viewed in three parts, of which the next two sections form the first. We strongly believe that the nature of the learning and instructional problems we hope to remedy with the aid of technology must be better understood if we are to be more successful. These two sections discuss the institutional context and students' backgrounds and beliefs concerning mathematics.

The following section provides a description of our efforts to use computers to supplement instruction in the various transition courses. It also hints at some of the directions in our current development and in our planning for the future.

The remaining sections indicate some of our current and future plans and include some reflections on potentially important uses of computers not currently implemented. I also include some warnings, especially concerning the debate about the effects computer algebra systems (now in hand–held form) should have on the curricula in algebra and elementary functions.

The Mathematical Sciences Teaching and Learning Center.

The Math Center was formed in 1980 in the College of Arts and Science and was supported in part by an NSF CAUSE grant from 1981 to 1985. It was established to support lower division mathematics instruction in the Department of Mathematical Sciences, serve as a base for in–service and pre–service development programs for secondary teachers, and conduct research on mathematics teaching and learning. Efforts to develop and evaluate uses of computer technologies to support these objectives have been a central interest from the beginning.

The Center's staff consists of two Ph.D. mathematicians, eight professional staff persons with advanced degrees in mathematics or physics, a professional programmer with an undergraduate degree in computer science, and several graduate and undergraduate student assistants.

The physical facilities of the Center consist of five rooms, including a classroom that contains a tutorial area and a computer area. The latter has 20 PLATO terminals and four IBM AT's with hard disks and printers. We plan to replace the terminals with a network of powerful microcomputer workstations by the fall of 1988. Much of the software for this new environment has already been developed; its features and purposes are discussed below. The design of the computing support and materials is guided by the Center's responsibility for teaching the large and heterogeneous populations of students enrolled in the transition courses.

The objectives, staff, and resources associated with the transition courses are designated the "Lower Division Program," which was formed in the Department of Mathematical Sciences and the Math Center in the fall of 1984. The forces that resulted in its creation were of two related forms: (a) central administrative concern about various complaints from some students and their parents concerning their experiences in mathematics, and (b) the continuing concern of both the faculty and the administration about poor performance of students in the University's mathematics courses.

The Math Center staff has responsibility for the following transition courses; both "traditional" and "self–paced/mastery" versions of each course are offered.

- *Intermediate Algebra* (M 010): A non–credit course designed for those not prepared for either of the following courses, as gauged by a placement test. Approximately 550 students take it each year. Its content is essentially a review of the topics in high school courses in introductory and intermediate algebra.

- *Elementary Mathematics and Statistics* (M 114): A three credit course taken by students from several professional colleges, such as nursing and agricultural sciences, but its largest constituency consists of students in the College of Arts and Science, who may use it to satisfy the mathematics requirement for the BA degree. Approximately 1000 students take the course each year. It consists of a review of algebra topics, followed by topics on elementary functions, then some trigonometry and a brief module on statistics.

- *Precalculus* (M 115): A three credit course taken by those in majors that require one or more semesters of calculus. It serves both the calculus sequence for business, biology, and selected other majors and the sequence for math, physics, engineering, and other natural science majors. Approximately 1300 students take the course each year. It contains relatively little explicit review of algebra, and it emphasizes the concept of function. The usual classes of elementary functions are treated.

The "traditional" course structure consists of two lecture– discussions per week for groups of 50 students. Each class is broken into two "workshops" of 25 that meet once each week. An instructional "unit" consists of these four class meetings, and the same instructor is responsible for all these meetings. All students in each course take the same examinations scheduled out–of–class in the evenings. Quizzes and other instructional resources available on computers are integrated into the instructional support for these courses.

The mastery/self–paced structure is essentially a modified Keller approach. Topics in each of the three courses are covered by cumulative mastery quizzes for each of twelve Instructional Units and a final examination. Students must take quizzes on an Instructional Unit until an 80% score is achieved before they can proceed to the next unit. Each student may "contract" for two or three credits.

These courses have been designed to meet the needs of a large and heterogeneous population of students. They are not yet models for the courses we believe should exist at precollege or college levels, however. We are not satisfied with some aspects of them, although the precalculus course does appear to prepare students reasonably well for the calculus course.

We tend to use the self–paced/mastery versions of courses to refine and experiment with tasks associated with the more demanding (and important) forms of understanding we seek. For example, there are a number of "themes" not effectively developed in most textbooks, such as "create" and "structure" tasks. The former require the student to create mathematical objects having certain attributes, e.g., to create a function having certain features in its graphic or analytic representations. (Examples appear later in Figure 8.) The latter require students to recognize structure in algebraic expressions (e.g., compositeness) and to use this information to construct or recall strategies or representations that might be useful for the task at hand. Such themes are supported by problems on virtually all the mastery tests.

Characteristics of Student Constituencies.

> "Mostly what I've been doing is just mainly from memory or its been trying to be from memory. I know there's a lot of rules in algebra that you have to learn first before you can go on to you know, that's the most important thing is to...learn those rules..." [4, p. 125]

This quote from a protocol for a student attempting to solve equations represents a common (and not too inaccurate) perspective on their algebraic experience. As currently taught, algebra and precalculus consist of an extraordinary number of apparently disaggregated topics and techniques. Virtually all textbooks reinforce this tendency through their organization into large numbers of three to five page sections, followed by long lists of relatively homogenous exercises on the newly developed topic. Learning is expected to occur through *examples* presented in the text and by the teacher. The more important inferences are often left to the student. Relatively little systematic attention is given to precisely those recurring patterns, strategies, and procedures that run through the mathematics curriculum.

We believe this confusion results, in part, from lack of clarity by mathematics educators and textbook authors concerning conceptual categories. For example, students should learn that nearly all the equations having closed form solutions in algebra and precalculus books must be transformed into one of three canonical patterns[1]:

- iterative use of disjunction,

$$f(X1)\,g(X2)=0 \;\leftrightarrow\; f(X1)=0 \text{ OR } g(X2)=0;$$

- decomposition,

$$f(X1)=f(X2) \;\leftrightarrow\; X1=X2; \text{ or}$$

- a set of polynomial methods (e.g., the quadratic formula).

We believe that knowledge of such categories can be an important unifying force in the curriculum. But examination of textbooks shows that these are left to be inferred by the student from an extraordinary array of *ad hoc* practice tasks.

The quote at the beginning of this section accurately reflects the attitudes of many of the students in our courses. They attempt to memorize or make rote application of vast amounts of low level information, such as all the individual steps in solving a rational equation. Then they attempt to match that pattern nearly symbol by symbol (when they can remember it) as they attempt a new problem that appears to be of similar type. This is, of course, an important initial phase of their mathematical development — but too often it ends there.

Many students depend almost entirely on this approach to mathematics. In fact, they believe this is what "understanding" mathematics means. Frequently they exhibit the following symptoms: They can do most of the 30 homework problems of similar type in a given section of the textbook on a given evening (or after briefly reviewing the steps in the pattern); they honestly believe they understand the topic or technique; and then they "crash" on the examination, which covers too many topics to permit most students, even those with above average memories, to use this approach successfully. They lament: "I could do every problem on the homework assignments — the problems on the test were different" (i.e., unrecognizable and therefore unfair).

The most powerful uses of computers to enhance mathematics learning will be those that "bridge the gap" between students' present beliefs, attitudes, and patterns of organization of their knowledge on the one hand and those needed for success in mathematics on the other. Hypotheses or assumptions or even "models" of both approaches are needed to guide effective uses of technology. For some illustrations of such hypotheses, see [24].

---

[1]For a fuller discussion of these methods in solving equations and an illustration of the contribution of artificial intelligence to mathematics education, see [18], [19], [20].

Modes of instructional uses of computers.

Several of the modes of use discussed here were developed initially in the early 1980's. Other colleges and universities have made similar efforts with some of them, although I suspect relatively few have had the resources to attempt quite so many. My purpose is not simply to report on this historical experience but rather to clarify the perspectives and modes that are evolving from this experience. The modes I discuss are:

- Practice with feedback;
- Diagnostics;
- Graphics environments;
- Tutorials in economics.

For each mode I discuss the pedagogical purposes, student constituencies served, forms of mathematics tasks, computer support, outcomes, and related forms under development or planned for the future.

Practice with feedback

Software named the Mathematics Interactive Problem Package (MIPP) has been developed in TUTOR for both instructional and research uses. The mainframe version currently runs on the University's PLATO system, and a microcomputer version now exists. The microcomputer version of this authoring system is written in C and runs on IBM PC's and compatibles. The screen displays that appear later in this paper (Figure 6) are from the microcomputer version.

MIPP consists of a Problem Database (multiple choice, true/false, and free response), a Problem Editor, a Curriculum Editor, Data Collection and Analysis features, Diagnostic Procedures, and other Miscellaneous Features.

Two different uses are made of our extensive databases of problems in algebra and elementary functions: practice with *mixed lists* of problems and quizzes. Approximately 1200 problems are available with their solutions provided in steps. Additional information about these tasks may be found in [23].

The database can be (and has been) rearranged according to whatever criteria can be used to classify the problems into lists. Typically headings similar to those used for chapters and subsections in textbooks are used (Figure 1). However, we believe that having one or more additional classifications available to the student is pedagogically important. The criteria used for the stratification of the problems (and, of course, the character of the problems themselves) are directly related to the kinds of associations we wish the learner to make among the concepts and procedures practiced. It is difficult to overemphasize the importance of this observation. Databases of carefully crafted math tasks

SCREEN 1:

a – Lessons to Use During the Orientation Period
b – Precalculus Diagnostics and Quizzes
c – Make Your Own Practice Problem Set
d – One–variable Function Plotter

SCREEN c:

(Partial List)

1	Functions	8	Exponential Functions
2	Function Operations	9	Logarithmic Functions
3	Inverse Functions	10	Exponential & Log Equations
4	Graphing Functions	11	Values of Trig Functions
5	Quadratic Functions	12	Graphs of Trig Functions
6	Polynomial Functions	13	Trigonometric Equations
7	Rational Functions	14	Trigonometric Identities

Figure 1. Screen displays listing resources for precalculus.

can play an important role in the development of associations and inferences that are too rarely formed from practice using the traditional classifications.

For the discussion that follows, however, the problems are organized in the traditional way. The software permits the student to select any set of topics (see Figure 1) and the total number of problems she or he wishes to see during the session. The problems then are presented in a random way. The student can see the solution to each as the problem is completed. The solutions are presented in "chunks." Each chunk typically represents the steps in achieving an appropriate subgoal of the problem. The student is urged to attempt to complete the problem after viewing as few steps as possible. Most of the problems are presented in multiple choice form, but the software supports several other forms of judging, including entering algebraic expressions.

For mixed lists of problems, the design of the presentation software was motivated by our desire to diminish some of the troublesome symptoms discussed in the previous section. We wanted students to discover *prior* to taking an examination that they were depending excessively upon "memory, match, and mimic" techniques, such as recalling what section of the text the problem is in, so they can list the strategies or procedures that might be useful.

At the time these databases were being formed, we did not yet have a perspective focused on the purposes of such practice. We hoped that such experience with mixed lists of problems might foster a more functional organization of their knowledge. But, as with the design of textbooks, the inferences we wished the student to make from such practice were still excessively *ad hoc.* We are currently devoting a good deal of attention to this issue. However, as the evaluation below indicates, even this relatively primitive instructional resource seems to help students who use it.

Weekly quizzes are a means of motivating students to do their homework regularly. They also provide feedback to students on their progress before the hour exam. But initially we created them quite grudgingly. We believe that some of the diagnostic uses discussed below have more pedagogical power than quizzes. But, to entice these students to use the wide array of instructional resources we have developed, we needed an incentive. Quizzes that provide a few points toward their course total (currently about 10%) were the initial solution to this troublesome problem.

We have developed software so an optical scanner can be used to score mastery tests and create appropriate student records from them. This capability also saves valuable tutorial time, as the items missed are immediately made explicit for the assistant, who reviews them with the student.

We are especially interested in using this capability to generate diagnostic messages for students and to refer them to additional instructional resources when appropriate. This capability can also be used in workshops associated with traditional forms of classroom instruction. For example, software programmed in Prolog is being prepared to search the database fields of a student, using whatever "IF–THEN" conditions we prescribe, and print appropriate messages. This greatly reduces the amount of computing equipment required to serve large numbers of students. However, as for other modes discussed, *the quality of such support is only as good as the quality of the tasks, their objectives, and the sequencing used.* The systematic development of helpful messages based on a student's behavior across an extensive variety of tasks is a very complex business if its objectives are to transcend the strictly procedural skills.

We also have extensive sets of problems and related instructional resources for the non–credit intermediate algebra course, M 010, using Control Data's PLM software. This software is a managed, objectives–driven structure. The course objectives and related mastery levels are set by the instructor, and the student is informed whether the objectives have been met. When they are not met, appropriate instructional resources are specified for the student's use. Use of this software will not be discussed further here.

Approximately 1200 students from M 115 and M 114 use the computer resources each week. In excess of 19,000 hours were logged by such users during the past calendar year. Statistics gathered to explore the effects of the use of the database of precalculus problems, the diagnostics, and the quizzes have been encouraging.

The data in Figures 2, 3, 4, and 5 provide an analysis of the performance of precalculus students in the fall of 1985 in relation to their use of quizzes, diagnostics, and practice with mixed lists of problems. The 787 students were broken into groups A, B, and C on the basis of their placement test scores (Figures 2 and 3). Each of these groups was separated into two (A1 and A2, etc.) according to the number of sessions they used the instructional computing materials (e.g., A1 used them > 10 times and A2 ≤ 10 times)[2]. Data on average VSAT, MSAT, combined SAT, high school percentile, algebra placement test score, and placement level were calculated for each group, most of which is shown in Figure 4. In each case the "1" groups (heavier users) performed significantly better in terms of total points on tests in the course and therefore received better final grades than the "2" groups, as shown in Figure 5. In fact, the B1 group performed significantly better than the A2 group, even though the A2 group had a mean MSAT of 592 and Algebra Placement Test score of 19.5, as compared with the B1 group's 550 and 15.3, respectively.

These data and positive results should be interpreted cautiously, however, since students who elected to make heavy use of these resources, while having weaker backgrounds, may be somewhat more motivated.

We believe that practice with a database of carefully selected problems can have an important influence on how students organize their knowledge of concepts and procedures. The criteria used for such organization must be carefully defined and support the associations, inferences, or uses we wish the learner to create. The microcomputer version of MIPP supports "levels" in its

[2]The number 10 was selected for the following reason: There are 7 quizzes and 7 diagnostics; most students take at least 6 quizzes, use a session for orientation, and experiment casually in one or more sessions. We were interested in the effects of the use of the mixed list of problems; those who signed on for more than 10 sessions were using that resource.

A: 17 < ALGebra placement test score
B: 13 < ALGebra placement test score ≤ 17
C: ALGebra placement test score ≤ 13

Figure 2. Student groups in precalculus.

PT Level	Algebra Score	Trig. Score	Enrollment Advice
I	0–8	–	M 010
II	9–10	–	M 010 or, if M 114 or M 115 is taken, participation in an algebra review program is recommended.
III	11–16	–	M 114 or M 115
IV	17–19	–	Participation in the precalculus review is strongly recommended if M 221, Calculus I, is taken.
V	20–30	0–3	M 210 or M 221[3]
VI	20–30	4–6	M 211 or M 241

(The total number of items on the placement test was 36.)

Figure 3. Placement levels, Fall 1985.

[3]M 210 is Introductory Discrete Mathematics, and M 211 is Introductory Calculus, normally taken simultaneously with M 210. M 221 is calculus for business, biology, and other majors; M 241 is calculus for mathematics, engineering, and physical science majors.

Group			MSAT	ALGEBRA PLACEMENT TEST	HIGH SCHOOL RANK	PLACEMENT LEVEL
A1	N = 25					
	MEAN	=	562	19.7	81	4.4
	S.D.	=	133	2.3	12	0.7
A2	N = 69					
	MEAN	=	592	19.5	79	4.4
	S.D.	=	65	1.5	14	0.7
B1	N = 106					
	MEAN	=	550	15.3	79	3.1
	S.D.	=	95	1.0	16	0.4
B2	N = 151					
	MEAN	=	542	15.4	77	3.2
	S.D.	=	84	1.0	16	0.4
C1	N = 155					
	MEAN	=	491	10.7	74	2.5
	S.D.	=	110	1.9	18	0.7
C2	N = 81					
	MEAN	=	512	10.6	74	2.4
	S.D.	=	76	1.9	16	0.7
LL	N = 787					
	MEAN	=	527	13.2	76	2.9
	S.D.	=	94	3.6	16	0.9

Figure 4. Previous mathematics preparation.

curriculum editors; these will be used in the future to experiment with this idea.

Diagnostics

Nine "Diagnostic Problem Sets" have been produced for precalculus students. A Diagnostic[4] consists of 8 to 15 problems, selected on the basis of students' behavior on precalculus examinations during the last four years. For most of these problems the student may enter algebraic expressions as answers. Each problem is designed to diagnose several different forms of conceptual difficulty or errors. For each such form,

[4]More extensive discussion of the diagnostic uses of computers, the negative side effects of textbook problems, and related learning research issues may be found in [23] and [24].

Group	N	A	B	C	D	F
A1	25	36.0	52.0	12.0	–	–
A2	69	36.2	23.2	26.1	7.2	7.2
B1	106	38.7	28.3	25.5	5.7	1.9
B2	151	20.1	24.5	24.5	17.9	12.6
C1	155	13.5	23.9	27.1	19.4	16.1
C2	281	6.8	17.1	26.7	19.9	29.5
TOTAL	787	18.6	30.0	25.7	15.8	17.0

Figure 5. Grade distributions of groups (percentages).

there is an interactive message designed to help the student identify his or her particular difficulty, to remedy it, and to practice other problems to avoid it in the future.

PROBLEM SIMULATION (press ESC to leave)

Which of the following is equivalent to $(10^{2n-1})^2(10^{1-n})^3$?

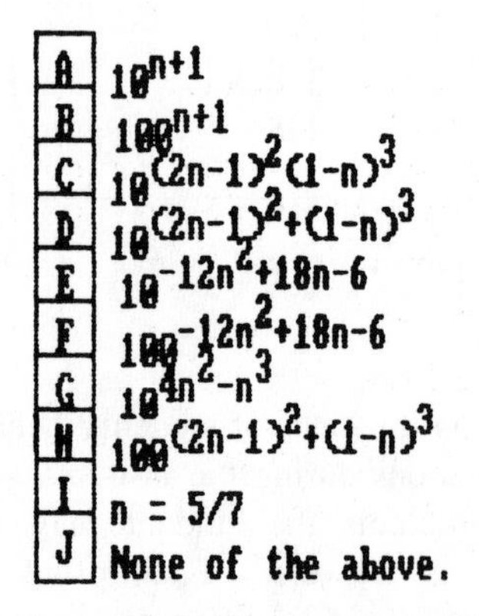

Press a letter to mark your answer.
F2-feedback/F5-correct/F6-solution steps

Figure 6a. A diagnostic item in MIPP.

It is generally essential that the written work of a significant number of students be studied carefully to develop a good diagnostic item, even one concerned with procedural skills. An example of such a problem is provided in Figure 6 (part a) together with two of the diagnostic messages (parts b and c).

In spite of the good results reported above, a pedagogical problem of some complexity had to be dealt with in order to encourage students to use these Diagnostic Problem Sets. College freshmen in the precalculus course (and other transition courses) are not as enthusiastic about understanding mathematics or even remedying poor algebraic habits as they are in getting a requirement out of the way. Thus, while the Diagnostic Problem Sets appear to be reasonably effective in helping students identify some of their difficulties, motivating students to use them voluntarily posed a problem. To encourage use, we built the on–line quizzes into the grading system for the course, but we required (in 1983) that students take the diagnostic before they were eligible for a quiz. For items on each Diagnostic, the student was asked (on–line) whether the message received was helpful. Approximately 85% of the answers were "yes." We used the data from the other cases to determine when a message required revision.

Our efforts to design diagnostic items and strands were frustrated by a very basic problem. A design that focuses on individual procedural errors of students

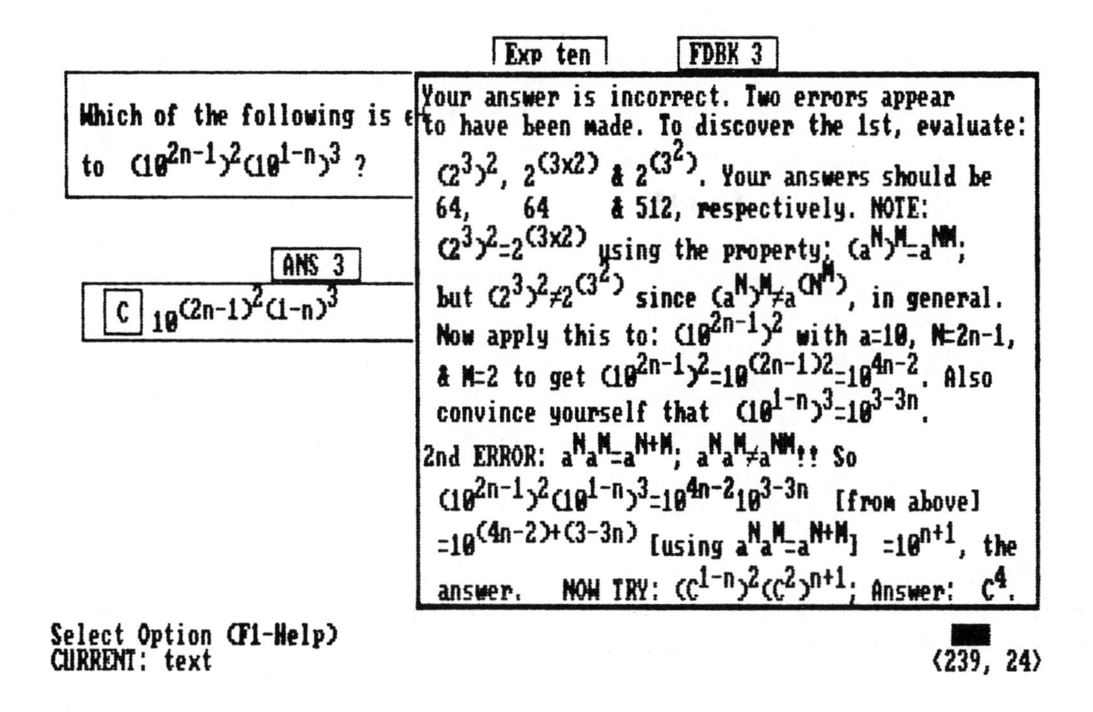

Figure 6b. A diagnostic response to an error.

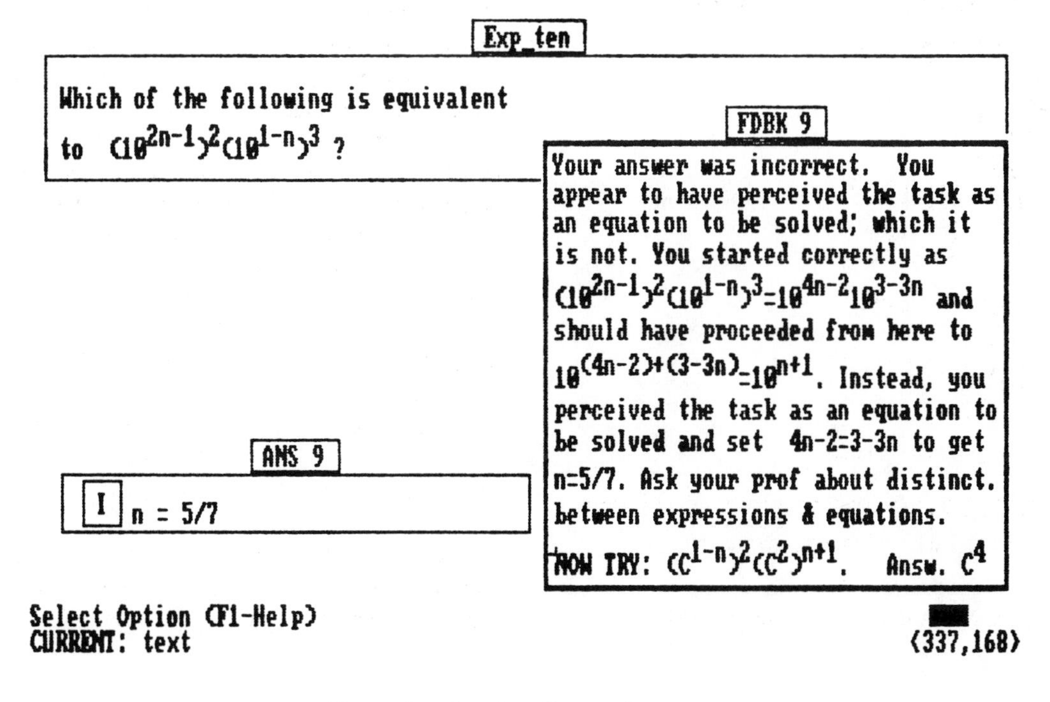

Figure 6c. Another diagnostic response.

invariably degenerates into excessive emphasis on an *ad hoc* set of procedural topics — a serious weakness in the existing curriculum. We believe the diagnostic emphasis must focus not only on the procedural level but on "strategic"[5] and conceptual levels as well. In particular, the complicity of the practice problems in textbooks in causing some of these difficulties deserves additional attention. For example, consider the ability to see composite patterns in the structure of algebraic expressions, such as

$$sin^2(3u) + 3\ sin\ (3u) + 2,$$

or for dealing with the chain rule in calculus.

One might want to consider answer options B through H for the problem in Figure 6a to be diagnosing the *same* basic collection of difficulties at one of two quite different levels.[6] The more basic level involves the lack of sufficient knowledge of the specific "rewrite rules" (such as $a^M a^N = a^{M+N}$) that apply to exponential expressions. However, many students know such rules but are unable to determine what is salient in them for a given context; e.g., in Figure 6a the exponents are algebraic expressions. Since there is a new set of rewrite rules for each new class of functions (e.g., logarithmic or trigonometric identities), it may be possible to focus on a diagnostic level that involves tasks designed to help students learn more generally how to interpret and use rewrite rules in algebra.

We believe that the effort to use computers for serious diagnostic purposes in mathematics can be shifted to a higher and more appropriate level by this work. For more details, see [1], [9], [10], [22], [23], and [24].

These Diagnostics were used with several hundred students during each of three semesters. A formative evaluation of them was conducted in precalculus in the spring of 1983 and was repeated in 1984 with similar results. Even though the diagnostics are somewhat primitive, the effects on students' performance in the course appear positive. For example, students who took no diagnostics had higher placement test scores than those who used them systematically (i.e., took more than six of the nine available), but the performance of the latter group was significantly better. At that time the

[5]For a concrete example of the meaning of "strategic," see the analysis in [24] of the behaviors of precalculus students when presented with the problem:

Solve $v\sqrt{u} = 1 + 2v\sqrt{1 + u}$ *for* v.

[6]Note that answer "I" involves a very different form of misunderstanding; see the message in Figure 6c. Five of 35 students taking a test on which this problem appeared viewed it this way.

Quizzes Attempted	No. of Students	% Passing on Test Pts.	Avg. Placement Test Score
$n > 6$	24	79.2	12.6
$5 \le n \le 6$	34	82.4	12.7
$3 \le n \le 4$	48	56.3	12.1
$1 \le n \le 2$	48	37.5	12.0
$n = 0$	84	42.9	14.9

Figure 7. Percentages passing Precalculus with quiz points excluded.

course was graded on a pass/fail basis, and students were not required to take the diagnostics. Figure 7 provides the percentages of students who accumulated the points needed to pass (excluding quiz points), stratified by the number of diagnostic/quiz pairs they elected to take. The figure also provides the average placement test scores for each group. (Motivation to take the diagnostics is likely to bias these results, of course.)

Our plans for diagnostic development are to:

a) prepare prototype tasks to monitor some important but selected errors in strategy and certain troublesome forms of conceptual misunderstanding;

b) implement these prototypes using our microcomputer MIPP authoring software.

Graphics environments

Mathematicians and mathematics educators do not need to be convinced of the importance of graphic representations of functions and relationships; their value is an unquestioned article of faith. Nor do we need to be convinced that computers might serve as powerful means both for using graphics representations for solving problems and for enhancing students' knowledge of graphs. How many function plotters have been written by the combined memberships of the MAA and NCTM?

However, an examination of a large number of textbooks in algebra and elementary functions reveals that, in spite of their emphasis on methods for graphing functions, relatively little *use* is made of such knowledge to solve other problems in the same books. Therefore, students do not tend to use graphic methods to solve problems unless explicitly told to do so.

Graphing becomes just another technique, like changing an algebraic representation by factoring. Thus, while we are all convinced that "thinking graphically" and "thinking to think graphically" are extremely important forms of behavior, we have simply not developed sufficient instructional materials and related practice tasks to develop such behavior.

At Delaware we have developed another attitude about the role of computer graphics environments in transition courses, namely, that tasks using such tools can also be used to teach concepts, notations, strategies, and even procedures that are not explicitly (sometimes not even implicitly) graphic. A primary theme in our effort to describe the forms of understanding we want to develop is that of change of representation, such as the usual one of [algebraic descriptions of functions] $\longleftrightarrow$ [graphic description], and the correspondence between features of one and of the other.

Certainly this theme is developed to a modest extent in most of our textbooks; we believe there is significantly more potential here. For example, relatively few books attempt to explore in a comprehensive way the relationship between literals (parameters) in algebraic expressions and families of graphs. Many do this tangentially in one or two contexts, but more systematic development of this and similar forms of correspondence may help students assign more meaning to such mathematical objects as abstract algebraic expressions than they currently do.

We are using tasks (see Figure 8) in our graphics environments to facilitate development of certain forms of understanding structure in algebraic expressions such as compositeness, correspondence between algebraic features and graphic features, and uses of functional notation when explicit algebraic referents are not given. We believe such graphic tasks might also be used to help diagnose and remediate some of the common algebraic errors students make, many of which are related to their lack of understanding of structure in algebraic expression (e.g., order of operations). Such errors are extremely difficult to remediate when the student's only representation of the phenomena is the symbolic one.

A last, and perhaps the most important, illustration of the variety of uses we are making of our graphics environment concerns our effort to influence students' beliefs about the meaning of "doing mathematics." Most of our students have had three to four years of high school mathematics. As already noted, many of them view "doing mathematics" as essentially a "memorize, match, and mimic" activity. Many are able to "SOLVE" a given equation or "GRAPH" a given function, but these same students initially have great difficulty with what we have come to call "CREATE" tasks: CREATE an equation having specified solutions; or, CREATE a function whose graph might match a given graph.

Such tasks require them to access and then organize their knowledge somewhat differently as they search for possible solutions. Some of the Graphics Lab Tasks assigned in precalculus are quite exploratory, open ended ones, planned so that the sequence of tasks imposes fewer and fewer explicit constraints.

1. Graph any three functions A, B and C you choose such that $C(x) < B(x) < A(x)$ for all real x. You may use functions from any of the classes we've studied or composite functions, if you wish.[7]

2. Create a quartic function with no x–intercepts and which has a minimum value.

3. Create an exponential function A such that $A(x) > B(x)$ for all x if $B(x) = \frac{x^3}{10} + 2$. Record your definition of A below and submit a graph of A and B plotted on the same coordinate axes.

4. Graph three linear functions all having positive slope and which form a triangle that contains the origin. Make a printed copy of your graphs and submit it with the lab task. Record the equations of your three linear functions below.

5. Graph three different lines with x–intercept at (0,–2). Record the equations of your three lines below. Write an equation representing any (and all) members of the "family" of lines with x–intercept (0,–2).

6. Graph the functions $A(x) = 2^x$ and $B(x) = \log_2(x)$. Graph a linear function C which passes through the y–intercept of function A and the x–intercept of the function B. You should use the cursor to determine the coordinates of these intercepts. Write the expression for the line below.[8] $C(x) =$

Figure 8. Examples of graphic lab tasks.

We also hope that tasks in which the student is encouraged to create or experiment with mathematical objects and concepts will lead to a different attitude about what mathematics is. There is some anecdotal evidence that this is occurring for some students, but those of us who attempt to alter habits, beliefs, and attitudes of thousands of freshmen students each year do not underestimate the enormity of this task.

The Graphic Lab Tasks have been used in lieu of seven quizzes assigned to sets of students in a precalculus course for two semesters. We plan to revise the current versions of the tasks into a "laboratory manual," which will be a central feature of the precalculus course. A version will also be prepared for the non–credit intermediate algebra course. Currently, students complete one set of tasks in each of seven weeks of the semester. Themes such as "structure in expressions," "families of functions," and "create tasks" are carried through several of the sets in a planned way, and the degree to which these exploratory tasks are constrained is varied in a planned way. Students' conclusions are recorded either by printing functions and graphs they have created or by writing their conclusions on the paper provided. They may take as long as they wish to complete the tasks, but they normally use approximately one hour. This work is evaluated by the instructor.

A very important dynamic in these sessions is that the students usually know (i.e., have graphic evidence) whether they have succeeded with the task. As a result, they appear to spend more time experimenting than do students who take quizzes from the database. We believe there is significantly more learning taking place during these sessions.

The graphic software was designed with some specific instructional objectives in mind, among them having students explore the graphic implications of structure in algebraic expressions and the description of such using functional notation. Thus, functions to be plotted may be written in terms of other functions either as composition or binary combinations (see Figure 9). Normal mathematics syntax is used rather than computer syntax to make it as easy as possible for the student to focus on the mathematics tasks and not be distracted by the interface. Scaling can, of course, be changed.

Evaluation of the tasks and this mode of computer use is underway; the initial information is very encouraging. Two instructors used these tasks during the spring of 1987 in precalculus. Each of the lecture sections (approximately 50 students) is split into two workshops; each instructor had one workshop group (randomly selected) use the Graphic Lab Tasks and the other use the computer database for quizzes. The two groups had essentially the same mathematics preparation in terms of placement test and MSAT scores.

[7]This task was given at the end of the semester. See Figure 10 for examples of student responses to it.

[8]This task was given in the first set, just after linear functions were reviewed, but before exponential and logarithmic functions were introduced.

CHANGE FUNCTIONS TO PLOT

Type the letter of the function to change.

EXPRESSION FOR FUNCTION A

$x^2 - 16$

EXPRESSION FOR FUNCTION B

A(x-2)

EXPRESSION FOR FUNCTION C

B(A(x))+A(x)

Or press: F1 to plot these functions
F2 for help
F4 to change parameters
F10 to return to the index

Figure 9a. Function entry screen for Graphics Environment.

The graphics groups scored more than 10% higher (more than 3 items of 30) on the final examination, had 20% higher mean scores on 14 of the 30 items, and had higher mean scores on 27 of the 30 items on the examination. Since there were other instructors and other workshop groups that did not use the graphics environment, the final examination was not inadvertently oriented towards the skills learned using the graphics tasks. In addition, the computer quizzes taken by the control group tend to consist of items of the type included on examinations. Thus, it appears that the topics covered in precalculus were better understood by the graphics groups *even though most of the problems were not explicitly graphic in character.*

To study the effects of this form of instruction more thoroughly, one of the instructors is:

a) collecting and evaluating videotaped traces of students' interactions with the computer on the various tasks;

b) examining their written work and printed records of graphs;

c) comparing their performance on the common out–of–class hour examinations and cumulative final examinations with that of the students who took multiple choice quizzes from our database;

d) developing a framework of themes and a more systematic description of the forms of understanding and inferences we believe such experience will help develop;

e) refining or designing the tasks relevant to learning those forms and inferences;

ONE-VARIABLE FUNCTION PLOTTER

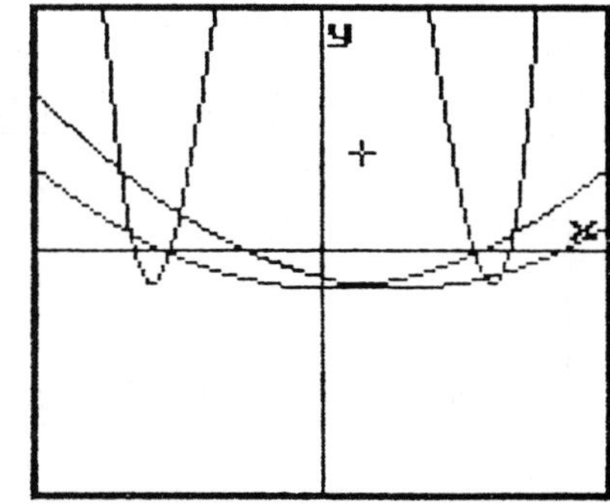

x_{max} = 7.00
y_{max} = 100.00
CURSOR
COORDINATES
x = 0.98
y = 40.00

Use arrow keys or
shifted arrow keys
to move the cursor.

FUNCTION A:

$x^2 - 16$

Press: SPACE for other functions
'?' for more options

Figure 9b. Display of the functions in Figure 9a.

Graph any three functions A(x), B(x) and C(x) you choose such that C(x)<B(x)<A(x) for all x. You may use functions from any of the classes we've studied or composite functions if you wish.

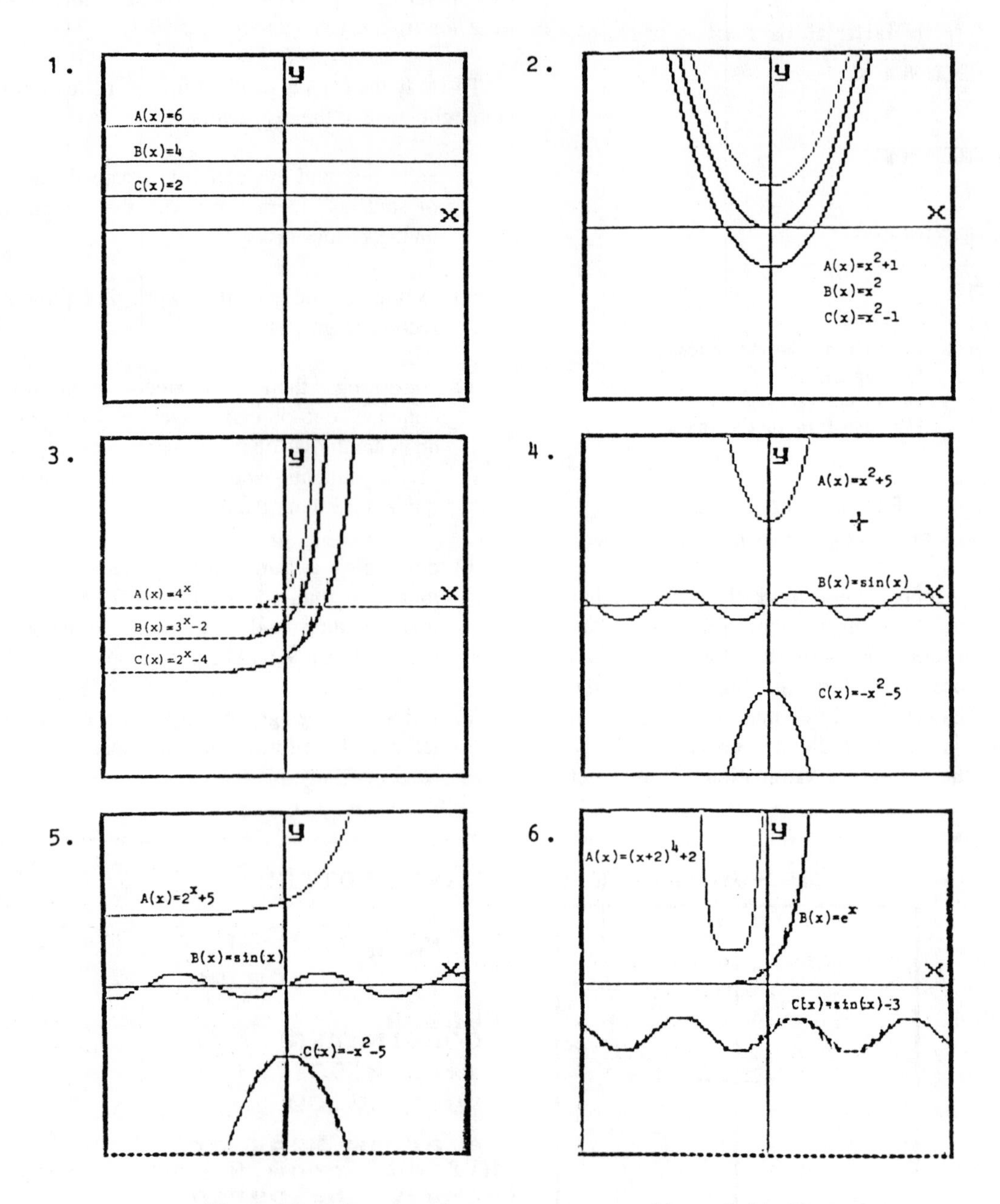

Figure 10. Selected student responses to graphics task 1 in Figure 8.

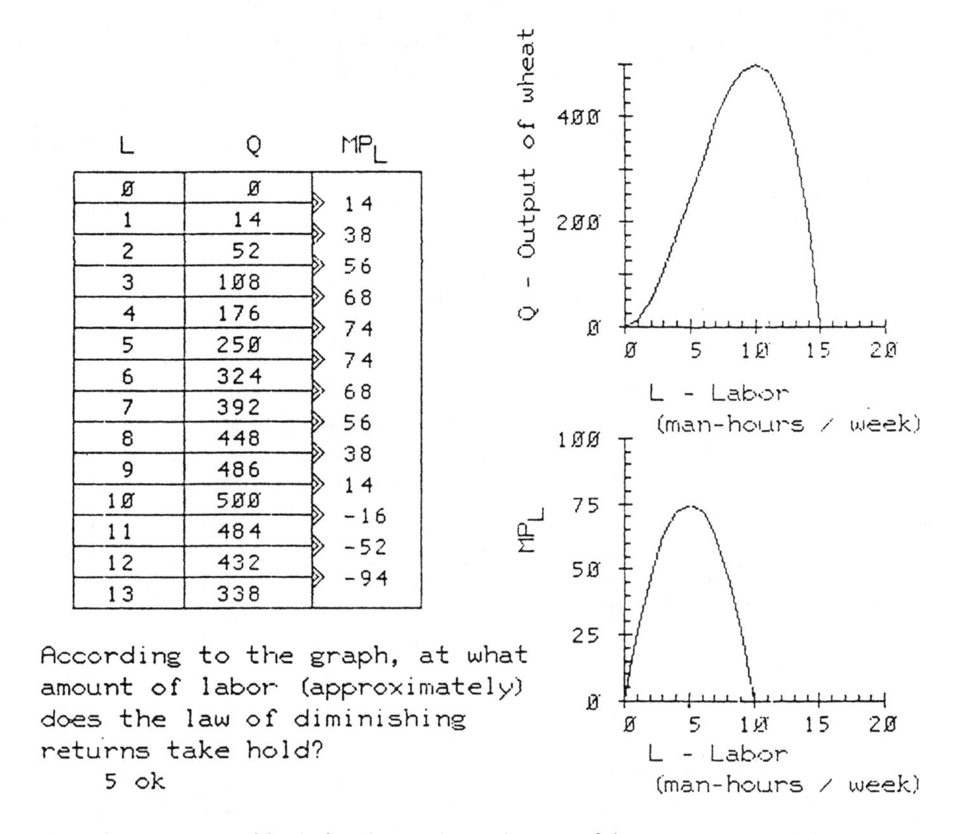

L	Q	MP_L
0	0	14
1	14	38
2	52	56
3	108	68
4	176	74
5	250	74
6	324	68
7	392	56
8	448	38
9	486	14
10	500	-16
11	484	-52
12	432	-94
13	338	

Figure 11. Display from an economics tutorial.

f) attempting to determine changes in attitudes, behavior, and beliefs concerning "doing mathematics."

We believe we can accumulate and interpret very useful information about these phenomena. There are methodological difficulties in measuring effects of this form of instruction, but we are convinced that students preparing to study calculus will benefit from such experiences. In fact, the work associated with d) and e) is essentially an effort to make more explicit the kinds of insight, skill, and inferences we know our students will need, but which are often too vague and implicit in practice tasks to expect them to be developed.

Tutorials in economics

The purposes of our precalculus/economics modules are to use important and troublesome economics topics to improve the student's grasp of the precalculus material and provide greater intuition about the concepts developed in the subsequent calculus courses. We want students to see how precalculus topics are used to model phenomena in a field in which they may have greater intrinsic interest.

The general pattern of each of the six lessons is to provide an informal interactive introduction to the central theme using a setting of interest to the student. This is followed by a formal development of the relevant concepts from mathematics and economics, again stressing interaction with the student. For example, the first three economics lessons are:

- linear functions in the economic context of consumption functions and the marginal propensity to consume;
- piecewise linear functions, quadratic functions, inverse functions, and concavity in

the context of production possibility curves and scarcity of resources;

- slope of a tangent line and its relationship with max/min and inflection points in the economic context of total, average, and marginal product of labor.

The remaining lessons take up cost functions, profit maximization, and supply and demand. These lessons were developed in TUTOR for use on the University's instructional computing system. The authors have developed other lessons with less demanding mathematics content. This software runs on the IBM PC and compatible equipment and accompanies the textbook [6].

The lessons are approximately 40 minutes in length. They are interactive, and every attempt has been made to use the computer's capabilities for *dynamic* graphics and for branching for wrong answer tutorials. They emphasize both conceptual and formal mathematical forms of understanding. Finally, the lessons provide constant reinforcement of the modelling process with questions designed to emphasize the two–way interaction between concepts and quantitative description. A sample of a screen display is provided in Figure 11.

The economics modules were evaluated by usage data, student evaluations, and faculty reports. They were first used in sections of precalculus and economics during the fall semester of 1983. Since that time they have been used in Economics 101 (Introductory Macroeconomics) and/or Economics 102 (Principles of Microeconomics) by their developers, but relatively little in precalculus, because it is difficult to integrate such extensive applications into the math course. The potential value of this tutorial mode of use justifies significant additional experimentation by the math community, however.

The informal student evaluations suggested that students found these tutorials useful but demanding. They were designed at least as much for their mathematics content as for their economics content. Thus, most economics students find their mathematical emphases very challenging. Although some economics faculty want their students to have just such mathematical sophistication, the department is unable to require it. For this reason, the use of the modules has been voluntary.

One of the economics faculty reports that students found the lessons in combination with the lectures to be extremely useful. He reports on his students' work on topics which are notoriously difficult to teach that

> "Performance, especially in the spring semester, was particularly good as a whole. These concepts were generally understood by students."

Research on learning.

There are a few references that provide an introduction to the research themes intertwined with our efforts to improve the effectiveness of the mathematics curriculum. Schoenfeld's book [17] is a good place to start. Others that are relevant include references [2], [3], [4], [7], [14], [15], [16], [18], [19], [20], [21].

I have found the perspectives from cognitive science very useful. This area is a blend of cognitive psychology and artificial intelligence. A reference I recommend to those writing textbooks in algebra and elementary functions is [20]. This book was an outgrowth of Silver's dissertation work at the University of Edinburgh, and it discusses the design of the Prolog program he developed to have the computer learn new solution methods for solving algebraic equations at the A–level on tests for university admission in Great Britain. The insight into patterns in algebraic expressions and their uses that such work provides has very interesting and useful implications for the teaching and learning of such knowledge. This work is discussed in more detail in [17] and [24].

Plans for the future.

The sections describing each of the modes of computer use contain information about current and future work. The primary projects and plans are summarized here.

A. The DIAGNOSTICS on microcomputer MIPP, using the optical scanner to database capability, will be developed. The emphasis in tasks will be extended to strategies and concepts.

B. The Graphic Lab Tasks are being revised into a laboratory manual.

C. We will develop an experimental interactive computing environment that will help learners reflect upon the organization of their knowledge and develop categories for such organization similar to those used by mathematicians.

D. We will continue developing more principled ways to describe the forms of understanding desired in terms that imply appropriate practice tasks in algebra and elementary functions.

The computer software that will be used to support our efforts is of five types: a) scanner ⟶ database; b) the graphics software; c) MIPP on microcomputers; d) the software that will be needed to support plan C above (this is being conceptualized but is not yet programmed); and e) Prolog programs that operate "above" the software in a) and c) to model the current state of development of the student with respect to selected math themes.

What forms of "understanding" do we want, anyway?

One of the primary reasons we find that so many remedial programs in mathematics are so unsuccessful is that we have not carefully analyzed the forms of mathematical behavior we expect. Without this, we are simply going by instinct and tradition. That such tradition has held up so long is an important index to its value — but we now need to be much more analytical about what we are trying to teach, especially if we are to use technology effectively. In particular, we must attempt to identify and define the forms of mathematical understanding we consider important and work backwards to curriculum and materials that serve them.

The answer to the question heading this section is: We don't know! This was acknowledged by mathematicians at a conference/workshop on the first two years of college mathematics [11]:

> "First of all, while there was considerable agreement at the conference that we should already do away with most *arithmetic* drill and allow students to use today's calculators instead, in contrast there was considerable worry about what doing away with algebraic drill would lead to. ... It may just be that skill and training in *symbolic* manipulation is closely tied to being successful as a mathematician, scientist, or engineer, or even to being an astutely analytic businessman; ... Clearly much more needs to be known about this issue."

> "We know what arithmetic skills we still want people to have in the computer age. They should know at least how to estimate effectively. That is, they still need good number sense. But what is the equivalent "good algebraic sense"? We don't know. Possible partial answers which were tendered: the ability to sense how many solutions a system of equations should have, and the ability to know what form an algebraic expression would best be worked into in order to draw from it easily whatever information is needed. But until we have a better idea of what algebraic sense people should have, we should be careful about throwing out the algebraic training we give now."

This statement represents the general perspective of some very able mathematicians on how little clarity exists concerning the forms of algebraic knowledge we desire. (It also implicitly assumes virtually no connection between arithmetic knowledge and algebraic understanding.)

Failure is virtually assured for many students if we cannot specify more clearly the forms of algebraic understanding we expect them to develop. There is little evidence that the kinds of understanding we wish to teach our students will form from the debris of topics, *ad hoc* techniques, concepts, and drills organized into small sequential sets in textbooks, unless the student is helped to develop dynamic and mathematically powerful categories for organizing such abstract symbolic chaos. This disaggregated condition is a result, to a large extent, of the lack of clarity concerning the desired forms of understanding.

Put another way, the current "bottom–up" approach through drill on many tasks must be guided and complemented by the identification of the algebraic structure, relationships, and skills that we identify as important. Methodologies useful for such identification include task analysis applied initially to some of the "higher level" tasks in the current curriculum [24]. Such "top–down" analyses are very effective ways of identifying important, but implicit, forms of understanding that may be taught more explicitly and, perhaps, more effectively and efficiently. The National Assessment panel indicated that "we cannot hope to improve substantially higher–level cognitive skills and understanding simply by teaching students lower–level knowledge and skills. ... Changes in higher cognitive levels will occur only when higher–level cognitive activity becomes a curricular and instructional focus" [13; p. 3].

The dynamic character of instructional computer environments is often more effective than are the more static, sequential, text–based media in focusing and *re*focusing the learner's attention on (a) the salient features of mathematical ideas and objects, and (b) the forming of associations with previous concepts or procedures required to support the inferences we expect them to make.

The computer algebra system (CAS) scenario.

What are the implications for the mathematics curriculum of a hand–held computer for symbolic algebraic manipulation, such as the Hewlett–Packard HP–28C? More specifically, there are two important questions for mathematics educators:

> What forms of skill and understanding are needed to use such tools in intelligent ways? (For example, what skill when using a CAS is analogous to estimation when using a calculator?)
>
> How can CAS environments be used in planned ways to teach and learn mathematics at the level of algebra and elementary functions?[9]

If a CAS can carry out many of the algorithmic tasks we currently spend so much time teaching students (often unsuccessfully) from middle school through lower division college, why teach them? This question is being raised frequently now by those interested in what the content of the mathematics curriculum should be in a technological world [12]. John Hosack provides some useful initial insights on this important theme in his paper in this volume.

Fortunately, even those most critical of the existing curriculum and excited about the new software have learned to be cautious about aggressively advocating that we throw out large sets of traditional tasks involving algebraic technique. Such caution is justified, in part, by the controversial history of the "new math" curriculum reforms and by the relatively modest impact the presence of calculators has had upon the elementary, middle, and high school curricula.

Caution is appropriate for at least two reasons. First, the objectives of the mathematics curriculum are a function of the constituencies being served. It will be excellent if we can effectively teach more students to develop models, to use mathematics utilities on computers (e.g., for linear programming) *with understanding*, and to analyze the models, without necessarily requiring them to master all the underlying mathematics. It will be even more helpful if such resources can be used to motivate students to want to study the underlying mathematics. But if the curriculum of the future *requires* a forced choice of academic track early in middle or high school, with serious implications for access to the mathematical training needed for scientific and engineering careers, we mathematics educators and society at large have serious problems to face.

The second caution concerns how little we currently know about how much concrete experience with algebraic symbols, relationships, procedures, and concepts individuals require to develop the mental representations needed for understanding the higher level concepts and structures normally associated with expert mathematical behavior. The issue was summarized in the Maurer quote in the previous section.

The roles of technology.

I believe that technology will come to play a number of important roles in the teaching and learning of mathematics. Computing environments in which powerful utilities (e.g., linear programming, curve fitting, computer algebra systems) are available, together with carefully designed tasks, exercises, and experiences, will have interesting impact on the curriculum. Virtually all current uses of technology are relatively *ad hoc* and still quite primitive. There is great oversell at the moment, but I expect significant improvement in the modes of use, the level of mathematics learning objectives served, and the quality of their implementation.

A factor seriously limiting robust and effective uses of computers to enhance mathematics learning is the lack of useful research on how individuals learn

---

[9]Other papers in this volume discuss the use of several CAS's as problem solving tools, but none of these CAS's is designed as an instructional environment for the level discussed here. One such will exist in the near future. It has been developed by WICAT Systems, and it supports a full algebra course. It is expected to be marketed soon by a major computer vendor.

mathematics — but this is changing. Efforts to use computers (e.g., as "intelligent tutors" in the sense of Sleeman and Brown [21]) are having strong catalytic effects on research on mathematics learning. I see many relationships between this research and our work with the development of diagnostic materials in precalculus and intermediate algebra. The complexity of many of the more powerful pedagogical ideas related to diagnostic materials *requires* technology.

Placement testing should now evolve into diagnostic testing, which would then have direct implications for the kinds of instructional materials needed. But to develop effective diagnostic materials that are not simply the aggregation of hundreds of our favorite problem types is the real challenge.

Conclusions.

I am somewhat cynical about the probability of success in *re*teaching the same topics in algebra and elementary functions in the same ways using the same text materials the students have seen in high school. I am encouraged by the increase in careful attention to and in the development of methodologies for studying our students' difficulties in mathematics more systematically. I am optimistic that some important and useful approaches will result. Some of these will require the use of computers and related interactive technologies. But I expect significant alteration in the design and objectives of text–based materials for mathematics during the next fifteen years — even those that have no computer software associated with them.

I believe that the most important effects of technology on the precollege and lower division college curricula will be its catalytic effects on rethinking what it is we are trying to do. Although the presence of computer algebra systems is one such catalyst (perhaps the most visible one, as articles in this volume suggest), I doubt that they will be any more influential in this respect than some of the perspectives and methodologies from cognitive science, when the latter are understood and used by mathematicians in the design of instructional materials.

I hope that some of our experience, perspectives and opinions will be helpful to others. I am especially pleased that our work in this area has led to very strong collaboration between math faculty at the University and our colleagues teaching mathematics in the high schools. We are also quite interested in collaboration with colleagues in secondary schools, community colleges, and universities on some of the development activities described in this paper.

We are optimistic that interactive computer technologies can play a very powerful role in teaching and learning mathematics. But we also have great respect for the tremendous amount of work required by mathematics educators to clarify their purposes so that such optimism may be realized.

Acknowledgments.

The Math Center was funded in part from July 1981 to May 1985 by NSF grant SER 8103901 under the Comprehensive Assistance to Undergraduate Sciences Education (CAUSE) program. The project title was "A comprehensive plan to imporve understanding and uses of the mathematical sciences by students in the State of Delaware using contemporary research on learning and educational technology." Other programs of the Center have been funded by the Greater Wilmington Development Council.

I wish to acknowledge the help of Math Center colleagues T. Ciro, B. Daley, M. Dolgas, B. Duch, J. Manon, B. Risacher, T. Seraphin, and J. Spamer. In particular, B. Duch first developed the mastery/self–paced approach for M 010 and M 114, and B. Daley extended it to M 115. J. Manon and M. Dolgas were the initial users of the graphics environments in M 115, and J. Manon is conducting the further study of the effects of these environments.

The graphics software was developed by M. Brooks (Mathematical Sciences), who also designed the MIPP software. The scanner ⟷ database software was developed by him and K. Hastings (also Mathematical Sciences), and its design was reviewed by B. Daley and B. Duch. The authors of the precalculus/economics modules were J. Bergman (Mathematical Sciences), C. Link and J. Miller (Economics).

REFERENCES

1. Booth, L. R. **Algebra: Children's Strategies and Errors: A Report of the Strategies and Errors in Secondary Mathematics Project**. Windsor: The NFER–NELSON Publishing Company, 1984.

2. Bundy, Alan. "A Treatise on Elementary Equation Solving." University of Edinburgh,

Department of Artificial Intelligence, DAI Working Paper No. 51, 1979.

3. Bundy, Alan. **The Computer Modelling of Mathematical Reasoning**. London: Academic Press, 1983.

4. Carry, L. R., C. Lewis, and J. E. Bernard. **Psychology of Equation Solving: An Information Processing Study**. Dept. of Curriculum and Instruction, University of Texas at Austin, 1981.

5. Davis, R. B. **Learning Mathematics: The Cognitive Science Approach to Mathematics Education**. London: Croom Helm, 1984.

6. Dolan, Edwin G. **Microeconomics** (Fourth Ed.). Hinsdale, IL: The Dryden Press, 1986.

7. Lewis, C. "Skill in Algebra." In John R. Anderson, ed., **Cognitive Skills and Their Acquisition**. Hillsdale, NJ: Lawrence Erlbaum Associates, 1981.

8. Lochhead, J., and J. Clement, eds. **Cognitive Process Instruction**. Philadelphia: The Franklin Institute Press, 1979.

9. Matz, M. "Towards a Process Model for High School Algebra Errors." In reference [21], pp. 25–50.

10. Matz, M. "Towards a Computational Model of Algebraic Competence." Master of Science Thesis, M.I.T., 1980.

11. Maurer, S. B. "The Effects of a New College Mathematics Curriculum on High School Mathematics." In A. Ralston and G. S. Young (eds.), **The Future of College Mathematics: Proceedings of a Conference/Workshop in the First Two Years of College Mathematics**. New York: Springer–Verlag, 1983.

12. National Council of Teachers of Mathematics Report, "The Impact of Computing Technology on School Mathematics." Reston, VA: National Council of Teachers of Mathematics, 1984.

13. **National Mathematics Assessment: Results, Trends and Issues**. Education Commission of the States, Report No. 13–MA–01, 1983.

14. Neves, D. M. "Learning Procedures from Examples." Ph.D. thesis, Department of Psychology, Carnegie–Mellon University, 1978.

15. Resnick, L. B., and W. W. Ford. **The Psychology of Mathematics for Instruction**. Hillsdale, NJ: Lawrence Erlbaum Associates, 1981.

16. Rissland, E. L. "Artificial Intelligence and the Learning of Mathematics: A Tutorial Sampling." In E. A. Silver (ed.), **Teaching and Learning Mathematical Problem Solving: Multiple Research Perspectives**. Hillsdale, NJ: Lawrence Erlbaum Associates, 1985.

17. Schoenfeld, A. **Cognitive Science and Mathematics Education**. Hillsdale, NJ: Lawrence Erlbaum Associates (in press).

18. Silver, B. "Using Meta–level Inference to Constrain Search and to Learn Strategies in Equation Solving." Ph.D. dissertation, Department of Artificial Intelligence, University of Edinburgh, 1984.

19. Silver, B. "Learning Equation Solving Methods from Worked Examples." Pp. 99–104 in R. S. Michalski (ed.), **Proceedings of the International Machine Learning Workshop**, University of Illinois, June, 1983. Also available from the Department of Artificial Intelligence, University of Edinburgh, as Research Paper 188, 1984.

20. Silver, B. **Meta–Level Inference**. Amsterdam: North Holland, 1986.

21. Sleeman, D. H., and J. S. Brown (eds.). **Intelligent Tutoring Systems**. London: Academic Press, 1982.

22. Wagner, S., S. L. Rachlin, and R. J. Jensen. "Algebra Learning Project: Final Report."

Report on NIE Project 400–81–0028. The University of Georgia, Athens, Georgia, 1984.

23. Wenger, R., and M. Brooks. "Diagnostic Uses of Computers in Precalculus Mathematics and Their Implications for Instruction." Pp. 217–231 in V. P. Hansen (ed.), **The 1984 Yearbook of the National Council of Teachers of Mathematics, Computers in Mathematics Education**. Reston, VA: National Council of Teachers of Mathematics.

24. Wenger, R. "Cognitive Science and Algebra Learning". In reference [17].

This book was edited and printed using the T^3 Scientific Word Processing System version 2.2. T^3 is a word processing system for IBM personal computers and compatibles. T^3 was designed by mathematicians for mathematicians.

Early drafts of this book were printed on dot matrix and PostScript laser printers. The final copy was printed on a Linotronic typesetter using the T^3 PostScript printer support.

The large headings are 17 point Times–Bold.

The medium headings are 12 point Times–Bold.

The main text is 10 point Times–Roman, 10 point Times–Bold and 10 point Times–Italic.

The gray bar at the top of most pages is a small graphics file included in the header named text.

For more information on T^3, please contact TCI Software Research, Inc., 1190–B Foster Road, Las Cruces, New Mexico, U.S.A., 88001. Telephone 1–800–874–2383 or 505–522–4600.